Lecture Notes in Computer Science   15229

Founding Editors

Gerhard Goos
Juris Hartmanis

Editorial Board Members

Elisa Bertino, *Purdue University, West Lafayette, IN, USA*
Wen Gao, *Peking University, Beijing, China*
Bernhard Steffen, *TU Dortmund University, Dortmund, Germany*
Moti Yung, *Columbia University, New York, NY, USA*

The series Lecture Notes in Computer Science (LNCS), including its subseries Lecture Notes in Artificial Intelligence (LNAI) and Lecture Notes in Bioinformatics (LNBI), has established itself as a medium for the publication of new developments in computer science and information technology research, teaching, and education.

LNCS enjoys close cooperation with the computer science R & D community, the series counts many renowned academics among its volume editors and paper authors, and collaborates with prestigious societies. Its mission is to serve this international community by providing an invaluable service, mainly focused on the publication of conference and workshop proceedings and postproceedings. LNCS commenced publication in 1973.

Henning Fernau · Inge Schwank ·
Jacqueline Staub
Editors

Creative Mathematical Sciences Communication

7th International Conference, CMSC 2024
Trier, Germany, October 7–10, 2024
Proceedings

Editors
Henning Fernau
University of Trier
Trier, Rheinland-Pfalz, Germany

Inge Schwank
University of Cologne
Cologne, Nordrhein-Westfalen, Germany

Jacqueline Staub
University of Trier
Trier, Rheinland-Pfalz, Germany

ISSN 0302-9743　　　　　　ISSN 1611-3349　(electronic)
Lecture Notes in Computer Science
ISBN 978-3-031-73256-0　　　ISBN 978-3-031-73257-7　(eBook)
https://doi.org/10.1007/978-3-031-73257-7

This work was supported by Austrian Science Fund (project 10.55776/P36688) and Vienna Science and Technology Fund (grant ICT19-065).

© The Editor(s) (if applicable) and The Author(s), under exclusive license to Springer Nature Switzerland AG 2025
Chapter "Large and Parallel Human Sorting Networks" is licensed under the terms of the Creative Commons Attribution 4.0 International License (http://creativecommons.org/licenses/by/4.0/). For further details see license information in the chapter.

This work is subject to copyright. All rights are solely and exclusively licensed by the Publisher, whether the whole or part of the material is concerned, specifically the rights of translation, reprinting, reuse of illustrations, recitation, broadcasting, reproduction on microfilms or in any other physical way, and transmission or information storage and retrieval, electronic adaptation, computer software, or by similar or dissimilar methodology now known or hereafter developed.
The use of general descriptive names, registered names, trademarks, service marks, etc. in this publication does not imply, even in the absence of a specific statement, that such names are exempt from the relevant protective laws and regulations and therefore free for general use.
The publisher, the authors and the editors are safe to assume that the advice and information in this book are believed to be true and accurate at the date of publication. Neither the publisher nor the authors or the editors give a warranty, expressed or implied, with respect to the material contained herein or for any errors or omissions that may have been made. The publisher remains neutral with regard to jurisdictional claims in published maps and institutional affiliations.

This Springer imprint is published by the registered company Springer Nature Switzerland AG
The registered company address is: Gewerbestrasse 11, 6330 Cham, Switzerland

If disposing of this product, please recycle the paper.

Preface

This volume contains the papers presented at the 7th Edition of the Creative Mathematical Sciences Communication (CMSC 2024) conference. The conference was held at the University of Trier from October 7–10. This conference marks another milestone in our ongoing journey to enhance computational thinking through innovative, constructionist, and hands-on activities. Since its inception more than a decade ago, CMSC has been a platform for fostering creativity and collaboration in the realm of computer science and mathematics.

Historical Context

The CMSC conference has a rich history, with previous editions held in esteemed locations such as Bergen (Norway, 2022), Poznań (Poland, 2020), Wellington (New Zealand, 2018), Lübeck (Germany, 2016), Chennai (India, 2014), commencing in Darwin (Australia, 2013), initiated by Michael R. Fellows and Frances Rosamond who were working there a decade ago. Over the years, the conference has taken on a biennial rhythm. Each of these gatherings has contributed to the growth and evolution of our community, expanding our understanding and approach to teaching computational thinking. This initiative can be traced back to the advent of "Computer Science Unplugged," about which we report below.

Conference Mission

Our mission is to address both the foundational elements and the contemporary, unsolved problems that challenge scientists today. The CMSC community upholds the belief that effective outreach is a two-way street. We recognize that listening to students enriches our scientific perspective, and, as such, previous editions have included visits to rural and Aboriginal schools and participation in science festivals.

New Directions in CMSC 2024

CMSC 2024 embraces two new and exciting directions: First, in contrast to earlier editions of CMSC, we introduced conference proceedings (more on this later). Second, on the final day of the conference, we engaged with the local educational community through the "Trierer Tag des Informatikunterrichts," an annual event for regional teachers interested in integrating computational thinking into their teaching. This initiative reflects our commitment to empowering educators and exploring innovative methods for fostering computational thinking skills, as exemplified by the CS Unplugged initiative.

About the CS Unplugged Initiative

CS Unplugged is an educational initiative that introduces fundamental concepts of computer science through engaging, hands-on activities that do not require the use of computers. The core philosophy of this movement is to emphasize that computers are merely tools, much like telescopes in astronomy, which aid in exploring larger, more intriguing problems. The true focus is on understanding and solving these problems algorithmically. By delving into the essence of computational thinking and problem-solving strategies without the need for digital devices, CS Unplugged aims to foster a deeper comprehension of the science behind computing and inspire curiosity and innovation in learners of all ages.

CS Unplugged was initiated by Tim Bell and Ian Witten from New Zealand and by Michael R. Fellows, in those days and back again now, from Australia. It is heavily driven by the idea to teach computational thinking already to primary school children, but it went far beyond that. Whoever is interested in these initial ideas and the philosophy behind CS Unplugged should consult https://www.csunplugged.org/en/computational-thinking/. Apart from collecting new ideas and updates on the mentioned website, there is also a book containing 'classical material' of CS Unplugged activities, freely available under https://classic.csunplugged.org/documents/books/english/CSUnplugged_OS_2015_v3.1.pdf. To quote the motivation behind this movement from the introduction of this book:

We have found that many important concepts [of Computer Science] can be taught without using a computer—in fact, sometimes the computer is just a distraction from learning.

The collection of teaching materials has gone viral over the years. Translations of all activities mentioned in the book (or at least many parts of it) are available in about 30 languages. As this is happening without a central driving force, it can be seen as a genuine grassroot movement.

Awards Connected to CS Unplugged

All the success did not get completely unnoticed over the years. Apart from several grants that were obtained in connection with the CS Unplugged project, we want to highlight two personal achievements that concern two of the founders of the initiative:

- Michael R. Fellows was awarded the highest civilian honor of Australia (Order of Australia,Companion to the Queen (AC), an honor descended from and essentially equivalent to a UK Knighthood) not only for his outstanding and well-known foundational contributions to Parameterized Complexity, but also for his contribution to the Computer Science Unplugged project.
- Tim Bell received New Zealand's highest Royal Honours for promoting public understanding of science and computational thinking.

These prestigious awards not only recognize the individual contributions of the founders but also underscore the enduring impact of the CS Unplugged project within the global educational community.

Publication and Peer Review

A second notable addition to CMSC 2024 is the opportunity for participants to publish their research papers, experience reports, and activity descriptions. This initiative ensures that the insights and innovations shared during the conference are preserved and accessible to future generations.

We allowed different types of submissions (all lightweight double-blind):

- long papers,
- short papers, and
- poster proposals.

The last type of submission was handled differently in the sense that these proposals are not included in the proceedings. For the long and short papers, we took a 2-round approach, with two different deadlines. Papers submitted to the first round not only received a binary accept-reject decision, but some authors were invited to revise their manuscripts according to the reviewer comments and then resubmit the paper to be reviewed a second time. This clearly helped improve the quality of the manuscripts. For papers (only) submitted in the second round, this benefit of integrating reviewer comments and then resubmitting the manuscript for a second review was not possible: here, we had to stick with binary decisions. Adopting this strategy, we could finally accept 7 out of the 9 short or long papers submitted in the first round. From the (purely) second-round submissions of short or long papers, we could accept 8 out of 15 submissions. Each submission was reviewed by up to four Program Committee members for quality, originality, and relevance. In the reviewing process and also for preparing the final version of the proceedings, we were using EquinOCS.

Conference Scope

The scope of content for this edition is diverse, covering a wide range of topics including educational activities for various topics in Computer Science as well as in Mathematics, curricular decision-making processes, competitions, and teaching in conflict zones. The last topic is indeed a new addition compared to previous editions of CMSC, but unfortunately it has quite some relevance in our world. These contributions highlight the breadth and depth of our field, showcasing the innovative approaches being taken to address current educational challenges.

According to the Call for Papers, the main topics of the conference include:

- Teaching Computer Science in early-childhood, primary, secondary, and higher education,
- CS Unplugged,
- Computational Thinking/Algorithmic Thinking,
- Broadening engagement and diversity for the exact sciences,
- Teacher education in Mathematics and Computer Science,
- Teaching CS in relation with other subjects,
- Contests and competitions,
- Education for students with special needs,

- Linking mathematical and computational content for fundamental competency development,
- Education within resource-limited environments,
- Remote learning amidst the effects of conflict.

We extend our heartfelt gratitude to all the authors for their valuable contributions, and to the Program Committee for their diligent efforts and support. Their dedication was instrumental in ensuring the success of CMSC 2024.

We are very grateful that we could attract two very prominent researchers in the field as keynote speakers; summaries of their talks can also be found in these proceedings. You can read about

- *Games That Cannot Go on Forever! Active Participation in Research Is the Main Issue for Kids* (talk delivered by Michael R. Fellows), as well as about
- *Solving Bebras-like Tasks: Approaches for Concept Building* (talk delivered by Valentina Dagienė).

We hope that the proceedings of CMSC 2024 will inspire and inform educators, researchers, and practitioners in the field of computer science and mathematics, contributing to the ongoing development of creative and effective teaching methodologies.

August 2024

Inge Schwank
Henning Fernau
Jacqueline Staub

Organization

Conference Chairs

Henning Fernau	University of Trier, Germany
Inge Schwank	University of Cologne, Germany
Jacqueline Staub	University of Trier, Germany

Steering Committee

Frances Rosamond	Lebanese American University, Beirut Lebanon
Małgorzata Bednarska-Bzdęga	Adam Mickiewicz University Poznań, Poland
Tim Bell	University of Canterbury, New Zealand
Valentina Dagiene	Vilnius University, Lithuania
Michael Fellows	Lebanese American University, Beirut Lebanon
Jam Ramanujam	Institute of Mathematical Sciences, Chennai, India
Rüdiger Reischuk	University of Lübeck, Germany
Matt Skoss	Australian Association of Mathematics Teachers, Australia
Verena Specht-Ronique	Hessisches Landestheater Marburg, Germany
Brett Stephenson	University of Tasmania, Australia

Program Committee

Sebastian Berndt	Technische Hochschule Lübeck, Germany
Sarah Carruthers	Vancouver Island University, Canada
Valentina Dagiene	Vilnius University, Lithuania
Christian Datzko	Gymnasium Liestal, Switzerland
Jens Gallenbacher	University of Mainz, Germany
Juraj Hromkovič	ETH Zurich, Switzerland
Tobias Kohn	Karlsruhe Institute of Technology (KIT), Germany
Dennis Komm	ETH Zurich, Switzerland
Regula Lacher	ETH Zurich, Switzerland
Barnaby Martin	Durham University, United Kingdom
Valia Mitsou	IRIF, Univercité Paris Cité, France

Matthias Mnich — Hamburg University of Technology, Germany
Xavier Munoz — Universitat Politècnica de Catalunya, Spain
Jean-Philippe Pellet — University of Teacher Education, Lausanne, Switzerland
Paweł Perekietka — Museum of Mathematics, Poland
R. Ramanujam — Azim Premji University, Bengaluru, India
Rüdiger Reischuk — ITCS, Universität zu Lübeck, Germany
Michał Ren — Adam Mickiewicz University, Poznań, Poland
Adish Singla — Max Planck Institute for Software Systems, Germany
Matt Skoss — Northern Territory Dept of Education, Australia
Bernadette Spieler — Zurich University of Teacher Education, Switzerland
Ulrike Stege — University of Victoria, Canada
Torstein Strømme — University of Bergen, Norway
Robinson Thamburaj — Madras Christian College, India

Additional Reviewers

Dirk Schmerenbeck — University of Trier, Germany

Teacher Day Committee

Jens Gallenbacher — University of Mainz, Germany
Jacqueline Staub — University of Trier, Germany
Inge Schwank — University of Cologne, Germany
Dirk Schmerenbeck — University of Trier, Germany
Martina Landman — TU Wien, Austria
Lukas Lehner — TU Wien, Austria
Christophe Stammet — University of Luxembourg, Luxembourg
Juraj Hromkovic — ETH Zurich, Switzerland

Local Organizing Committee

Henning Fernau — University of Trier, Germany
Jacqueline Staub — University of Trier, Germany
Sabine Laros — University of Trier, Germany
Dirk Schmerenbeck — University of Trier, Germany
Esther Stürmer — University of Trier, Germany

Contents

Invited Papers

Solving Bebras-Like Tasks: Approaches for Concept Building 3
  Valentina Dagienė

Games That Cannot Go on Forever! Active Participation in Research Is
the Main Issue for Kids 15
  Michael R. Fellows and Frances A. Rosamond

Tactile Learning: Unplugged Graphs, Trees, and Patterns

Unplugging Dijkstra's Algorithm as a Mechanical Device 39
  Riko Jacob and Francesco Silvestri

Unplugged Decision Tree Learning – A Learning Activity for Machine
Learning Education in K-12 50
  Lukas Lehner and Martina Landman

Tactile Kolam Patterns – Communicating Art and Mathematics to Students
with Vision Impairments 66
  *Robinson Thamburaj, Krishnamachari Desikachari,
  and Gnanaraj Thomas*

Teaching Advanced Concepts Using Tangible Machines

QuBobs Teaching Kits to Explain Quantum Computing 77
  Sophie Laplante, Loris Perez, Sylvie Tissot, and Lou Vettier

Solid Geometry Modeling: 3D Printing Is Not Always the Best Option 93
  Matthias Müller, Benjamin Weißing, and Pascal Lütscher

The Algorithm Experience at Primary Schools: An Experience Report 104
  Maarten Löffler

Curricular Desicion-Making

Curricular Choices for Computational Thinking in Large Scale Low
Resource Environments 117
  R. Ramanujam and Vipul Shah

Why Teach About Binary Numbers? 130
  Tim Bell and Henry Hickman

Teaching Tangible Division Algorithms or Going from Concrete
to Abstractions in Math Education by the Genetic Socratic Method 136
  Juraj Hromkovič and Regula Lacher

Computational Thinking and Interdisciplinary Instruction

Mathematical *Versus* Computational Thinking with a Computer
in the Background 147
  Maciej M. Sysło

Computational Thinking Based STEM Art Exhibits 162
  Jay Thakkar and Manish Jain

BeLLE: Detecting National Differences in Computational Thinking
and Computer Science Through an International Challenge 168
  Heidi Kaarto, Javier Bilbao, Arnold Pears, Valentina Dagienė,
  Janica Kilpi, Marika Parviainen, Zsuzsa Pluhár, Yasemin Gülbahar,
  and Mikko-Jussi Laakso

Innovative Teaching Beyond the Classroom

From Caesar Shifts to Kid-Enigma. The CS Unplugged-Like Path
in the MuMa Science Centre 185
  Michał Ren, Paweł Perekietka, and Łukasz Nitschke

Large and Parallel Human Sorting Networks 194
  Stefan Szeider

Distance Teaching of Mathematical and Computer Disciplines During
the War in Ukraine 205
  Galina Bulanchuk, Oleh Bulanchuk, Olena Piatykop,
  and Valentyna Ilkevych

Author Index 215

Invited Papers

Solving Bebras-Like Tasks: Approaches for Concept Building

Valentina Dagienė[✉][iD]

Institute of Educational Sciences, Vilnius University, Vilnius, Lithuania
`valentina.dagiene@mif.vu.lt`

Abstract. When students start learning the basic concepts of computer science (CS), they quickly find opportunities to demonstrate their skills, share interests, and compare their work with others. Attraction, innovation, techniques, and surprise should be desirable features of each task presented to learners. This paper discusses Bebras-like tasks that are designed to convey basic CS concepts to all age of students. Bebras is an international initiative aimed at promoting informatics and computational thinking among K-12 students, as well as teachers. Solving these tasks can be considered a systematic process that involves students in a deeper understanding of computing concepts and supports a pedagogical shift in the classroom, fostering students' engagement and motivation to learn. Short tasks covering CS concepts can be solved in a few minutes and can be presented either on a computer or printed on cards in an attractive design. Another way to introduce Bebras-like problems is through unplugged activities, where a hands-on approach using physical objects, components, or their own bodies helps students to better grasp CS concepts.

Keywords: Computer Science Education · Computational Thinking · Bebras-like Task

1 Introduction

The Bebras Challenge on informatics and computational thinking is an educational community network that unites teachers and researchers dedicated to teaching informatics/computing/computer science (CS) in schools. Launched in Lithuania in 2004, the Bebras challenge has expanded to over 85 countries (Bebras, 2024), with about four million students participating in 2023–2024 alone. The goal of the Bebras challenge is to engage students in CS, enhance their understanding of computer technologies, and promote learning through the solving of brief, concept-based exercises and puzzles, known as Bebras tasks.

The Bebras challenge is designed to spark interest in CS among school students across various age groups: Pre-Primary (ages 5–8), Primary or Little Beavers (ages 8–10), Benjamin (ages 10–12), Cadets (ages 12–14), Juniors (ages 14–16), and Seniors (ages 16–19). Participants use a computer to solve 15 to 18 tasks of varying difficulty within a 40–45 min timeframe, with each task taking approximately 1 to 4 min to complete. These tasks come in different types: multiple-choice with four options (text, numbers or images), open-ended (digits, numbers, characters, words or phrases), and dynamic (different levels of interactivity).

The Bebras challenge aims to engage in CS-related activities and deepen students' comprehension of computers from a young age, including kindergarten, and to promote creative use of modern technologies in learning activities. The challenge emphasises the understanding of basic CS concepts over technical details, and prioritises the application and interpretation of results, the control of computations, and the conceptual understanding of computers. Cognitive, social, and cultural aspects play a significant role in the challenge, which highlights the importance of culture and language in the use of technology (Fischer, 2009; Dagienė & Futschek, 2013; Dagienė et. al., 2014).

The Bebras challenge is thoughtfully designed to promote fundamental CS concepts for both boys and girls, ensuring that its appeal is equally strong across genders. This approach has yielded positive results: significant participation from girls has been noted in recent contests, with some countries even achieving near parity in gender representation. The challenge serves as an inclusive platform for students of all grades to engage in problem-solving activities, with the primary focus during the annual Bebras Week (held in November) being on solving and discussing the tasks rather than on competition. Collaborative problem-solving is encouraged, with formats such as pair or team participation (as seen in Germany) being particularly welcomed.

The Bebras challenge not only fosters widespread participation but also ensures that its objectives are both engaging and accessible to all participants. By focusing on well-designed tasks that highlight fundamental concepts, the challenge makes a significant contribution to CS education globally.

2 Bebras Tasks

The challenge comprises a series of short questions or tasks, known as Bebras tasks, tailored to different age groups. These tasks can be answered without prior knowledge of computing but are clearly related to CS concepts. Attraction, invention, tricks, and surprise should be desirable features of each task presented to students. The Bebras tasks need to be carefully designed, taking into account CS education and the pedagogical value of each task, the CS concepts it teaches and its attractiveness to learners —whether it sparks their motivation to learn.

A concept is not a single piece of information. To grasp a concept, a student needs to create many isolated mental models of this concept. Isolated concepts allow learners to make connections in their minds. This implies a learner must go through a number of situations in which the concept appears through various prisms and from different viewpoints. To solve Bebras tasks students are required to think in and about information, discrete structures, computation, data processing, but they also must use algorithmic concepts and problem-solving skills. The short tasks are the starting point of the problem-solving process.

At the heart of the Bebras challenge are the tasks, which present CS concepts and phenomena in a brief, engaging format through creative stories, visuals, and interactive elements. The quality of these tasks is vital for sustaining the interest of both students and teachers. To successfully motivate students to learn CS and enhance their computational thinking skills, it is essential to effectively promote the challenge, actively involve schools and teachers, and ensure the inclusion of high-quality tasks. Additionally, these tasks should be engaging enough to encourage teachers to discuss them with students post-competition, integrating them into CS lessons to introduce new concepts or simplify complex topics.

Numerous studies delve into the exploration and analysis of Bebras tasks. For example the paper (Dagienė & Dolgopolovas, 2022) examines how short tasks within the global Bebras challenge support and enhance students' computational thinking skills. The authors argue that the Bebras challenge, with its series of carefully designed short problems, provides an effective framework for introducing and reinforcing computational concepts. By analysing the structure and impact of these tasks, the study highlights how they facilitate incremental learning and problem-solving skills in diverse educational settings. The findings suggest that these short tasks are instrumental in nurturing computational thinking by offering accessible, context-rich scenarios that challenge students to apply and refine their problem-solving abilities.

Another paper (Datzko & Datzko, 2021) investigates aspects of designing effective Bebras tasks and emphasizes that successful tasks should be both engaging and pedagogically sound. The study highlights the need for tasks to be age-appropriate, challenging but solvable, and relevant to the core CS concepts. Effective tasks often incorporate elements of gamification and problem-solving that capture students' interest and foster deep understanding.

The paper of Lonati (2020) presents a detailed analysis of how Italian teachers use Bebras tasks to inspire and develop computing topics in their classrooms. The study examines the adaptation and integration of these tasks into teaching practices, highlighting their role in deepening students' understanding of computing concepts. Through a series of interviews and observations, the author identifies the strategies teachers employ to make Bebras tasks engaging and effective for their students. The findings highlight that Bebras tasks serve as a significant motivational tool and offer a practical approach to teaching CS concepts, allowing for creative adaptation in different educational contexts. The study underscores the potential of Bebras tasks to support diverse instructional methods and to promote a deeper engagement of students with computing topics.

Bebras tasks, designed to be solved within minutes, are crafted to inspire, motivate, and engage students by incorporating elements of attraction, innovation, and surprise. These tasks can be tackled without prior computing knowledge but are intrinsically linked to CS concepts. Understanding these concepts requires students to build multiple mental models, connecting isolated ideas in their minds. This necessitates encountering the concepts in diverse contexts and from different viewpoints. Bebras-like tasks compel students to engage in critical thinking about information, discrete structures, computation, and data processing while utilizing algorithmic and problem-solving skills.

From the inception of the Bebras challenge, task developers have sought to create engaging and exciting task formulations, identifying and categorizing distinct task groups. Continuous consultations on relevant topics have been an integral part of this process. Principles regarding task structure and topic selection have been discussed extensively by researchers (Dagienė & Futschek, 2008; Opmanis et al., 2006; Dagiene &

Stupuriene, 2016a; Dagiene et al., 2019). The selection of topics for Bebras tasks is closely related to teaching the CS fundamentals. Consequently, the development of these tasks considers common educational elements in CS across various countries, particularly those active in the international Bebras community.

Bebras tasks are divided into three levels of complexity, each with different point values: 6 points for easy tasks, 9 points for medium tasks, and 12 points for hard tasks. Correct answers earn the designated points, while incorrect answers result in point deductions (-2, -3, and -4 points, respectively). Unanswered questions receive 0 points. To prevent negative scores, participants begin with a baseline score equal to the maximum possible subtractions (e.g., 54 points for 18 tasks).

The Bebras challenge employs several types of tasks to engage students, including interactive tasks, open-ended tasks, and multiple-choice tasks. These task types are designed to stimulate student interest and encourage deeper exploration of informatics concepts. Creating these tasks is a complex endeavour, challenging both researchers and educators to develop and present engaging and context-rich problems.

Multiple-choice tasks feature four well-defined answer options, with only one correct choice, aimed at assessing students' understanding of specific concepts. Interactive tasks require students to actively engage with the computer, involving actions such as dragging and dropping objects, clicking on images, manipulating items with the keyboard, and selecting elements from lists. This interactivity fosters a hands-on approach to problem-solving. The dynamic nature of these tasks is pivotal, as it involves a two-way exchange of information between the user and the system.

Below are several examples of Bebras tasks designed for different student age groups and educational levels illustrating typical Bebras tasks and the CS concepts they aim to teach. Each task encourages students to explore the targeted CS area or CS concept, interact with data, and apply knowledge related to fundamental CS topics. By offering various solution strategies, these tasks support diverse problem-solving approaches and cater to students of different ages and abilities.

2.1 Stamping (Lithuania, 2010 for 6–8 years Old)

There are five stamps numbered from 1 to 5 (Fig. 1a) and a picture stamped using these stamps (Fig. 1b).

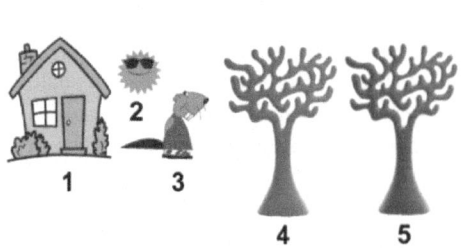 

Fig. 1. The five stamps (a) and the picture created using them (b)

In what order were the stamps used?

A) 5–2–4–3–1
B) 5–3–4–2–1
C) 5–2–3–4–1
D) 5–4–2–3–1

Explanation of CS concepts. The task requires figuring out the sequence in which five numbered stamps were used to create a picture. It introduces fundamental CS concepts such as coding basic actions as commands, sequencing commands, and even simple elements of reverse engineering. Through this activity, even younger children can develop insights into computational thinking. While the task itself does not involve formal algorithms, it helps children understand command prioritization and problem-solving strategies, making it a valuable exercise for fostering algorithmic thinking.

2.2 Jumping Kangaroo (Lithuania, 2020, for 10–12 Years Old)

A kangaroo jumps home. She can jump along the path, only vertically (up – down) or horizontally (left – right) and only if there are not more than two bricks in the way. The kangaroo wants to be home as quickly as possible (Fig. 2).

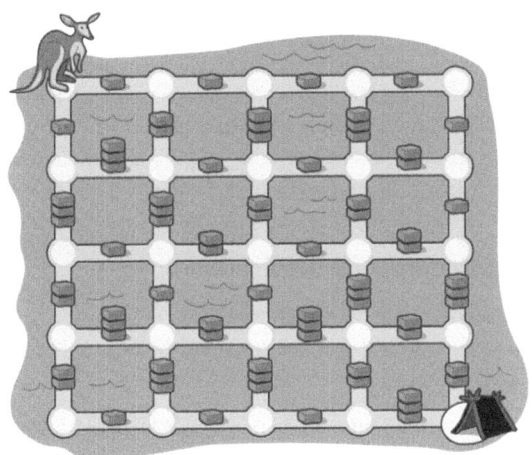

Fig. 2. The kangaroo path

How many jumps does the kangaroo need to make?

Explanation of CS concepts. To find the solution, follow these steps: Start by searching step-by-step. If you reach a dead end where all possible paths have three stones blocking the way, backtrack (possibly several steps) and try different paths. This method, known as *backtracking* in computer science, is a technique used in many algorithms to solve puzzles, Sudoku, or combinatorial optimization problems.

Sometimes, it is more efficient to start solving the problem from the end, working backwards from the kangaroo's home. In this case, doing so requires less backtracking, making the solution easier to find. However, without initially exploring the problem, it is impossible to determine whether it is better to start from the beginning or the end.

Solution: 14 jumps.

2.3 Nuts and Bolts (Canada, 2022, for 12–14 years Old)

At the Beaver Construction factory, Benoit works at the nuts and bolts assembly line (Fig. 3).

Fig. 3. A construction factory

Benoit's job description is as follows:

- He stands at one end of a long conveyor belt, which contains a line of nuts and bolts.
- His job is to take each element, either a nut or a bolt, off of the conveyor belt.
- If Benoit takes a nut from the conveyor belt, he puts it in the bucket beside him.
- If Benoit takes a bolt from the conveyor belt, he grabs a nut from the bucket beside him, attaches the nut and bolt together, and places the assembled part onto large box.

However, things can go wrong for Benoit in two different ways:

1. If he takes a bolt from the conveyor belt, and there is no nut in the bucket to attach.
2. If there are no more nuts or bolts on the conveyor belt, and there are still nuts in the bucket.

Which sequence of nuts and bolts, when processed from left-to-right, will **not** cause things to go wrong for Benoit?

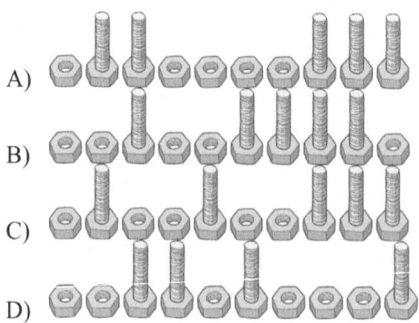

Explanation of CS concepts. This task, designed for more advanced students in the senior group, highlights the use of a push-down automaton (PDA). A PDA describes an algorithm that relies on the current state and has unlimited memory in the form of a stack. In this task, the state is either having a nut or a bolt on the conveyor belt, and the stack is the bucket that holds the nuts.

A PDA can be used to recognize or parse context-free languages. To recognize or parse a language means to determine if a given sequence of symbols belongs to that language. In this case, the nuts and bolts can be seen as a representation of balanced parentheses, where N = (and B =). Balanced parentheses are valid arrangements of parentheses in arithmetic expressions. Examples of unbalanced sequences include (((() or ()). Detecting balanced parentheses is crucial in compilers, as many programming languages use parentheses to indicate nested scopes and arithmetic expressions.

Developers of Bebras tasks strive to select intriguing and relevant problems to motivate students and enhance their understanding of technology. They aim to cover a broad range of CS and computer literacy topics. Despite the presence of educational standards for CS in some countries, there is no universal agreement on an integrated syllabus for CS education.

3 Concepts and Approaches for CS Education

The approaches to teaching CS in schools should be multidimensional, allowing students to see computers as more than just a collection of buttons and instructions. They should be understood as fundamental elements of cultural development and innovation. Introducing computing at an early age is crucial, as it equips students with essential problem-solving and critical thinking skills. Viewing computers as tools for generating ideas, rather than just finished products, enhances students' conceptual appreciation. This perspective fosters a deeper understanding of how CS principles integrate with broader cultural and intellectual contexts, preparing students for a future where technology plays an integral role in all aspects of life (Dagiene et al., 2017).

Bebras tasks play a pivotal role in introducing CS concepts. According to a commonly agreed-upon definition, each Bebras task should encompass at least one CS concept. These tasks are designed to focus on smaller learning items, making complex ideas more digestible for students (Dagiene, 2016; Dagiene & Stupuriene, 2016b).

The concept can be described as an extensive piece of information about a particular object perceived by the human senses. The content of a concept can vary significantly depending on personal experience. In CS, concepts are closely linked to our goals regarding the foundational knowledge we wish to impart to children and at what age. Deciding what to include in CS education for primary and secondary schools is challenging due to several factors (Dagienė & Stupurienė, 2016b):

Various approaches can be used to bring Bebras tasks to students, such as using online platforms like Moodle or ViLLE, employing cards with Bebras-like tasks, adapting tasks to unplugged activities, and incorporating games. These approaches facilitate a systematic process that leads to a deeper understanding of computing concepts, supporting a pedagogical shift in the classroom and enhancing student engagement and motivation. By focusing on microlearning, these short tasks facilitate learning in small, manageable increments. They can be presented digitally or on attractively designed cards, aligning with the hands-on approach many students need to effectively grasp complex concepts.

When designing the Bebras challenge, these principles were kept in mind. Short CS concept-based tasks are a powerful approach that can drive a pedagogical shift in the classroom, increasing students' engagement and motivation to learn. Such tasks encourage deeper understanding and critical thinking in CS. This approach can be a strategic tool for educators aiming to stimulate students' interest and promote a deeper learning experience. Bebras tasks can cover fundamental and advanced topics within CS, providing students with a broad and nuanced understanding of the field.

Using Bebras-like tasks, students can be introduced to various CS concepts:

- *Algorithms and programming*
- concepts, such as sequential and concurrent processes. These tasks might involve problem-solving scenarios where students need to design or analyse algorithms that operate in a specific sequence or handle multiple tasks simultaneously.
- *Data structures* are another key area that can be explored through Bebras tasks. Concepts like heaps, stacks, and queues can be incorporated to challenge students in organizing and manipulating data. For example, a task might require students to efficiently manage a list of items using a stack or to solve problems involving queues.
- *Modelling of states, control flow, and data flow* – these concepts are crucial for understanding how programs execute and how data is processed. Bebras tasks can involve scenarios where students must map out state transitions, control structures like loops and conditionals, or visualize data flow through a system.
- *Human-computer interaction* can be addressed by designing tasks that require students to consider user interface design, usability, and interaction principles. Such tasks might present students with a design challenge for a user-friendly interface or ask them to analyse how different interaction methods affect the user experience.

One more concept, an abstraction, is a cornerstone CS concept for developing students' ability to manage complexity and think at higher levels of problem-solving (Mirolo et al., 2022). It involves distilling complex systems into simpler, more manageable components, allowing learners to focus on essential features without getting bogged down in details. This skill is fundamental not only to writing efficient code but also to understanding and designing algorithms, data structures, and software systems. By mastering abstraction, students can better grasp how different parts of a program interact, leading to more effective debugging, optimization, and innovation. Ultimately, teaching abstraction empowers students to tackle complex real-world problems, fostering a deeper understanding of both theoretical and practical aspects of CS.

A study on comparison how different types of Bebras tasks—specifically those focused on abstraction and those on algorithmic thinking—affect students' understanding and skills in computational thinking (Vaníček et al., 2021) conducted. The findings reveal that both task types have distinct educational benefits, with abstraction tasks fostering conceptual understanding and algorithmic tasks enhancing problem-solving skills. The research contributes valuable insights into designing effective Bebras tasks and optimizing their use to support students' development in computational thinking.

Lehtimäki et al. (2022) focuses on developing and implementing a module that integrates Bebras tasks to teach computational thinking concepts. The module employs

Bebras-like tasks with the objective of providing engaging and practical experiences that assist learners in comprehending the fundamental concepts of computer science. The findings underscore the potential of Bebras-like tasks to improve computational thinking education and prepare both students and teachers for real-world problem-solving challenges.

Integration of the Bebras challenge within learning analytics-enriched environments was investigated by Pluhár et al. (2022). The study analysed how data from the Bebras challenge can be utilized to provide deeper insights into student performance and learning outcomes. The research suggests that learning analytics can offer valuable feedback for both educators and students, supporting more personalized and data-driven approaches to teaching computational thinking and problem-solving skills.

Solving Bebras tasks significantly enhances students' understanding and skills in programming, as evidenced by numerous studies and scientific papers (Araujo et al., 2017; Heintz & Mannila, 2018; Vinikienė et al., 2023; Vaníček et al., 2022; Vaníček et al., 2023). These tasks, designed to be engaging and educational, require students to apply fundamental programming concepts and problem-solving strategies in a practical context. By presenting programming concepts through problem-solving activities, the Bebras tasks foster computational thinking and provide a foundation for future coding skills.

One approach to using Bebras tasks is situated learning, which integrates these tasks into real-world contexts to enhance their relevance and impact. This method encourages students to engage with computing concepts through practical scenarios and problem-solving activities that reflect real-life applications. By embedding Bebras tasks into learning environments, educators can foster deeper comprehension and facilitate meaningful connections between theoretical knowledge and practical skills. This approach aims to enhance students' ability to apply computational thinking in various situations, fostering deeper learning and retention of CS concepts (Bellettini et al., 2019).

The importance of the CS Unplugged approach is becoming increasingly apparent, as evidenced by recent research (Bell & Vahrenhold, 2018; Bell and Lodi, 2019; Kuo & Hsu, 2020). Various CS Unplugged activities are connected to Bebras tasks, highlighting their educational value. It is crucial to understand that CS unplugged is not a standalone curriculum and is not intended to replace the opportunity for students to write programs on digital devices. Instead, it serves as an adjunct pedagogy that allows learners to grasp broader computing concepts without the immediate need to program. Additionally, it engages students in significant ideas through physical activities, reducing screen time and promoting an active learning environment.

CS concept-based tasks on cards, derived from Bebras tasks, support the CS Unplugged approach (Dagienė et al., 2019). These card-based tasks offer a more compact format, with metadata published separately either on additional cards or online. To facilitate this method, a set of cards featuring CS concept-based tasks has been developed. Each set includes an additional card that lists the informatics concepts along with links to the corresponding task cards.

The computational thinking board game "Robot City," included as an attachment to compulsory education learning materials in Taiwan, also utilized Bebras Challenge tasks in the pre-test and post-test phases of their empirical study (Kuo & Hsu, 2020).

This demonstrates the similarities between the computational thinking skills promoted by this unplugged game (e.g., a card-based board game) and those measured by Bebras Challenge. Therefore, in the near future, it is suggested that the international Bebras Challenge team collaborate with such an unplugged game to develop new Bebras Challenge tasks so as to extend both impacts.

Yan-Ming and Ju-Ling (2022) investigate how gamifying Bebras tasks can influence students' engagement and cognitive development in computational thinking. A game "Captain Bebras" is designed to present Bebras-like tasks in an interactive game format, aiming to make the learning process more engaging and effective. The findings suggest that incorporating Bebras challenges into digital games can improve students' computational thinking by offering a dynamic and enjoyable learning environment.

4 Conclusion

The goal of designing effective educational tasks, such as those in the Bebras challenge, is to transform mental models into coherent conceptual models. Traditional drill-and-practice methods may struggle in this regard, as they often lack the emotional engagement necessary for reshaping biased mental models. Conversely, a challenge-based environment offers an emotionally engaging context, promoting the integration of interdisciplinary concepts into a cohesive mindset.

Bebras education exemplifies a problem-solving approach, allowing personal engagement with the subject matter in a meaningful context. This approach facilitates socialization through an international problem-solving environment and fosters the "Aha!" moments crucial for learning. Effective task design should consider the analogue-digital duality, enabling the interplay of interdisciplinary concepts to trigger emotional and personally significant mental models while developing unbiased conceptual models. This process should be supported by scaffolding, teaching heuristics to replace biased mental models with corrected ones, and progressing towards a robust informatics-related conceptual model.

To effectively engage students and establish CS as a vital scientific discipline, it is essential to focus on creating meaningful and stimulating educational experiences. Successful involvement can be achieved through well-structured activities that feature intriguing and challenging tasks. By immersing students in the world of CS through such engaging tasks, educators can facilitate a deeper understanding of core concepts and foster a genuine interest in the field.

Furthermore, integrating a variety of pedagogical techniques—such as gamification, project-based learning, and collaborative problem-solving—can enhance students' learning experiences. Providing context-rich scenarios and gamification helps students appreciate the impact of CS on various aspects of society and technology. By creating an educational environment that is both engaging and reflective of the dynamic nature of informatics, educators can inspire students to pursue further studies and careers in this important field.

Acknowledgments. Bebras tasks (including graphics) are developed by Bebras challenge community based on Creative Common license CC BY-SA. I would like to thank Professor Ting-Chia Hsu from National Taiwan Normal University for reading the paper and providing valuable comments.

References

Araujo, A.L.S.O., Santos, J.S., Andrade, W.L., Guerrero, D. D.S., Dagienė, V.: Exploring computational thinking assessment in introductory programming courses. In: 2017 IEEE Frontiers in Education Conference (FIE), pp. 1–9. IEEE (2017)

Bebras – International Challenge on Informatics and Computational Thinking. Homepage: http://www.bebras.org. Accessed 03 Aug 2024

Bell, T., Lodi, M.: Constructing computational thinking without using computers. Constr. Found. **14**(3), 342–351 (2019)

Bell, T., & Vahrenhold, J.: CS unplugged—how is it used, and does it work? In: Adventures Between Lower Bounds and Higher Altitudes: Essays Dedicated to Juraj Hromkovič on the Occasion of his 60th Birthday, pp. 497–521 (2018)

Bellettini, C., Lonati, V., Monga, M., Morpurgo, A., Palazzolo, M.: Situated learning with Bebras tasklets. In: Pozdniakov, S., Dagienė, V. (eds.) Informatics in Schools. New Ideas in School Informatics, ISSEP 2019. LNCS, vol. 11913, pp. 225–239. Springer, Cham (2019). https://doi.org/10.1007/978-3-030-33759-9_18

Dagienė, V.: Bringing informatics concepts to children through solving short tasks. Bullet. EATCS **1**(118) (2016)

Dagiene, V., Dolgopolovas, V.: Short tasks for scaffolding computational thinking by the global Bebras challenge. Mathematics **10**(17), 3194 (2022)

Dagienė, V., Futschek, G.: Bebras international contest on informatics and computer literacy: criteria for good tasks. In Informatics Education-Supporting Computational Thinking: Third International Conference on Informatics in Secondary Schools-Evolution and Perspectives, ISSEP 2008 Torun Poland, 1–4 July 2008 Proceedings 3, pp. 19–30. Springer, Berlin Heidelberg (2008). https://doi.org/10.1007/978-3-540-69924-8_2

Dagiene, V., Futschek, G.: Bebras, a contest to motivate students to study computer science and develop computational thinking. In: learning while we are connected, vol. 3, book of abstracts, pp. 139–141 (2013)

Dagienė, V., Futschek, G., Stupurienė, G.: Creativity in solving short tasks for learning computational thinking. Constr. Found. **14**(3), 382–396 (2019)

Dagiene, V., Mannila, L., Poranen, T., Rolandsson, L., Stupuriene, G.: Reasoning on children's cognitive skills in an informatics contest: findings and discoveries from Finland, Lithuania, and Sweden. In: Informatics in Schools. Teaching and Learning Perspectives: 7th International Conference on Informatics in Schools: Situation, Evolution, and Perspectives, ISSEP 2014, Istanbul, Turkey, 22–25 September 2014. Proceedings 7 (pp. 66–77). Springer (2014). https://doi.org/10.1007/978-3-319-09958-3_7

Dagienė, V., Sentance, S., Stupurienė, G.: Developing a two-dimensional categorization system for educational tasks in informatics. Informatica **28**(1), 23–44 (2017)

Dagiene, V., Stupuriene, G.: Bebras–a sustainable community building model for the concept based learning of informatics and computational thinking. Inform. Educ. **15**(1), 25–44 (2016)

Dagiene, V., Stupuriene, G.: Informatics concepts and computational thinking in K-12 education: a Lithuanian perspective. J. Inf. Process. **24**(4), 732–739 (2016)

Datzko, C., Datzko, S.: Aspects of designing a successful bebras challenge. In: Conference ISSEP 2021 Online Local Proceedings (2021). https://www.ru.nl/publish/pages/1026345/datzko_aspects_of_designing_a_successful_bebras_challenge.pdf. Accessed 03 Aug 2024

Fischer, G.: End-user development and meta-design: foundations for cultures of participation. In: End-user Development, pp. 3–14. Springer, Berlin (2009)

Heintz, F., Mannila, L.: Computational thinking for all: an experience report on scaling up teaching computational thinking to all students in a major city in Sweden. ACM Inroads **9**(2), 65–71 (2018)

Kuo, W.C., Hsu, T.C.: Learning computational thinking without a computer: how computational participation happens in a computational thinking board game. Asia Pac. Educ. Res. **29**(1), 67–83 (2020)

Lehtimäki, T., Hamm, J., Mooney, A., Casey, K., Monahan, R., Naughton, T.J.: A computational thinking module for secondary students and pre-service teachers using Bebras-style tasks. In: Proceedings of the 2022 Conference on United Kingdom & Ireland Computing Education Research, p. 1 (2022)

Lonati, V.: Getting inspired by bebras tasks. How Italian teachers elaborate on computing topics. Inform. Educ. Int. J. **19**(4), 669–699 (2020)

Mirolo, C., Izu, C., Lonati, V., Scapin, E.: Abstraction in computer science education: an overview. Inform. Educ. **20**(4), 615–639 (2022)

Opmanis, M., Dagiene, V., Truu, A.: Task types at "Beaver" contests. Inf. Technol. Sch., 509–519 (2006)

Pluhár, Z., et al.: Bebras Challenge in a Learning Analytics Enriched Environment: Hungarian and Indian Cases. In: Bollin, A., Futschek, G. (eds.) Informatics in Schools. A Step Beyond Digital Education, ISSEP 2022. LNCS, vol. 13488, pp. 40–53. Springer, Cham (2022)

Vaníček, J., Šimandl, V., Dobiáš, V.: Bebras tasks based on assembling programming code. In: Bollin, A., Futschek, G. (eds.) Informatics in Schools. A Step Beyond Digital Education, ISSEP 2022. LNCS, vol. 13488, pp. 113–124. Springer, Cham (2022). https://doi.org/10.1007/978-3-031-15851-3_10

Vaníček, J., Šimandl, V., Klofáč, P.: A comparison of abstraction and algorithmic tasks used in Bebras challenge. Inform. Educ. **20**(4), 717 (2021)

Vinikienė, L., Dagienė, V., Stupurienė, G.: Introducing programming concepts through the bebras tasks in the primary education. In: Keane, T., Fluck, A.E. (eds.) Teaching Coding in K-12 Schools: Research and Application, pp. 145–156. Springer, Cham (2023) https://doi.org/10.1007/978-3-031-21970-2_10

Chen, Y.-M., Shih, J.-L.:. Bebras in the digital game "Captain Bebras" for students' computational thinking abilities. In: CTE-STEM 2022 conference (2022). https://proceedings.open.tudelft.nl/cte-stem2022/article/view/455. Accessed 03 Aug 2024

Games That Cannot Go on Forever! Active Participation in Research Is the Main Issue for Kids

Michael R. Fellows and Frances A. Rosamond

Lebanese American University, Beirut, Lebanon
michael.fellows@uib.no, frances.rosamond@lau.edu.lb

Abstract. It is entirely interesting, and profoundly important to science, that efforts to communicate science have often led scientists to new perspectives on their own work and their scientific fields and specialties. This two-way street in science communication was a founding impulse and inspiration of the Creative Mathematical Sciences Communication (CMSC) conference series. The vigor of that impulse continues in this paper via a few new game horizons. This paper uses well-quasi-ordering of trees and other mathematical objects to create games that are easy to understand, addictive, and engage some powerful mathematical fundamentals.

Keywords: Mathematical Sciences Communication · Games · Parameterized Complexity · Well-Quasi-Ordering · Artificial Intelligence

1 Introductory Remarks and Background

Providing new ways for children to authentically engage in science is foundational to the CMSC project. Outreach in science can be a *two-way street* best based in authentic research participation by children and the broader community. This paper describes several examples in the Introductory Remarks and Background, and the main part of the paper is concerned with an exciting class of mathematical games that comes from current research in algorithms and complexity.

Example 1. "This is Mathematics? This is MEGA-Mathematics!"
These were the words of a primary school boy who was seriously engrossed in trying to find a method to properly color a graph (network of dots and lines) with using only three colors. The rule for the coloring problem is that any two vertices connected by an edge must get different colors. He was passionately working and looked up at us to enthusiastically proclaim, "This is Mathematics? This is MEGA-Mathematics!" Yet, everyone knew what he would be when he grew up. He was known as the class criminal.

Why was he so excited about trying to find a clever and efficient method to color a graph using only three colors according to the one rule that two vertices connected by an edge must receive different colors? It is because nobody knows a fast method to solve this problem. There is a simple method of determining whether or not a graph can be colored with two colors with the same rule, and for very simple instances of the problem it can be easy to decide if three colors is enough. But as soon as the number of vertices in the graph increases beyond about 30, deciding if the graph can be properly colored with three colors becomes very difficult to solve and for large graphs it would take years to decide even on the world's fastest computer using the best algorithm currently known. Nobody knows whether a fast algorithm to determine if a graph can be 3-colored is possible. Currently, a fast algorithm for this NP-hard computational problem is not known. A million dollars is on offer as a reward to anyone who either: (a) finds a fast algorithm, or (b) proves mathematically that a fast algorithm cannot exist.

The young boy's exclamation became the title of a book and project of the Los Alamos National Laboratories in New Mexico [1]. The goal of the project was to bring the experience and content of 20th century mathematics to elementary school students and their teachers. The project was the favorite of the Director [2]. The book was a predecessor to the materials and book, *Computer Science Unplugged!* which has been used around the world and translated into over 35 languages [3].

The young boy's enthusiasm for this problem is not unique. The eyes of almost all children light up when they understand that some mathematical question is unsolved: *nobody knows!* We have received letters from children with suggested solutions long after they have heard about various open problems in one of our workshops. They have become authentically engaged in research, hooked because they are engaged with a question where *nobody knows* the answer.

There is a million-dollar reward by the Clay Foundation for an efficient method of determining whether a graph can or cannot be properly colored using only three colors or for a proof that an efficient algorithm for this computational problem is intrinsically impossible. This is called the "P versus NP" conundrum [4]. When children hear this, their amazed response is, "Even for children?" The answer is, "Yes, math is fair. The reward is for anyone who settles the issue."

Example 2. The Degree/Diameter Problem
The authors spray-painted on the sidewalk the largest known planar (no edges crossing) graph where every vertex has maximum degree three (at most three edges touching the vertex) and diameter three for a third grade workshop. Diameter three means that it takes only three hops along an edge (line) to go from any vertex to any other vertex.

Children were busily engaged in picking a place (vertex) to stand on. Then they would pick another place at random and try to figure out how to get there in three hops. They were hopping around on this spray-painted planar graph.

The teacher was wondering what this had to do with mathematics. The logic that the children were learning is used in every calculus class. They were learning alternating logical quantifiers: "...*for every node u* in the graph and *for every other node v, there exists* a path from u to v of length 3. The mathematical concept of "for every... there exists" is absolutely fundamental.

The graph painted on the sidewalk was a planar graph and had 12 vertices [1]. Is it possible to design a graph that has maximum degree three and diameter three and has more than 12 vertices? *Nobody knows.* They found this then-unsolved problem addictively fascinating and exciting.

The daughter of one of the authors came home from school saying, "School is boring. Well, not the degree-diameter problem. Me and my friends meet at lunchtime to try to solve it. Well, not all lunchtimes and not every recess. Tommy thought he had a new solution with 14 vertices. He made a mistake. There were two nodes with no length 3 route between them in his graph."

"Why isn't this problem boring?"

"Because nobody knows the answer and maybe we can figure it out."

This child was speaking like a true mathematician. What research colleague would enthusiastically agree to work on a problem that had already been solved? They would look at you like you are crazy. Posing questions that young people can understand and that haven't been solved is along the lines of teaching methods such as those in the movie *Radical*, based on the true story of teacher Sergio Juarez Correa at the Jose Urbina Lopez Primary School in Matamoros, Mexico [5] (Fig. 1).

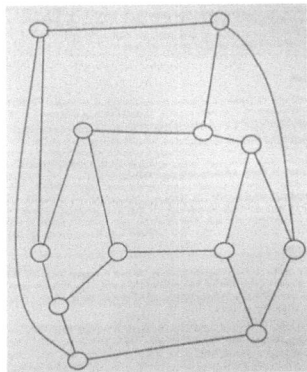

Fig. 1. The largest known planar graph with maximum vertex degree 3 and diameter 3.

Example 3. Kid Krypto

Historically, it is entirely interesting, and profoundly important to science, that efforts to communicate science have often led scientists to new perspectives on their own work and their scientific fields and specialties. This is the two-way street. This has certainly been the case for theoretical computer science, where the effort to communicate the basic idea of public key cryptography led to a major new branch of basic research in this area. This two-way street in science communication was a founding impulse and inspiration of the CMSC conference series [6].

The Kid Krypto encryption system demonstrates the mathematics underlying public-key cryptography without using advanced mathematics [3, 7]. Another Kid Krypto-style encryption system that is based on disjoint cycles in a graph or network is also easily accessible to a very young audience [8].

These combinatorics-based PKCSs (the technical name is public key crypto system) are easily accessible to primary-level schoolchildren, illustrating key mathematical ideas central to modern civilization. After working on Kid Krypto with Neil Koblitz and creating this enticing Kid Krypto system based on perfect codes in graphs that has proven a powerful motivation for children, the authors realized that their questions and thinking about the communication that had taken place with children could turn Kid Krypto into a whole new branch of "adult" cryptography. Through the work of many, this became a branch of modern cryptography (based on polynomial Groebner bases), of keen current interest because such combinatorics-based cryptosystems might be immune to attack by quantum computing [9].

Example 4. The Sorting Network
The Sorting Network is one of the original *Computer Science Unplugged!* activities and now an iconic activity used in workshops and classrooms around the world to introduce the concept of an algorithm and parallel processing. Children walk along the edges from vertex to vertex (comparator nodes) acting as computer processors working in parallel. Further discussion of the Sorting Network and other off-line activities can be found in the book, *Computer Science Unplugged!* by Tim Bell, Ian Witten and Mike Fellows which is available for free at www.csunplugged.org [3].

A history of the *Computer Science Unplugged!* project is given in [10] (Fig. 2).

Fig. 2. Children on a spray-painted Sorting Network

Computers frequently *sort* input values, which in computer science usually refers to putting a list into alphabetic or numerical order (phone numbers, customers, etc.). There are several sorting methods, such as selection sort, quicksort, bubble sort, insertion sort, merge sort and others. For a given problem, it is important to choose a method that is fast and does not use too many computing resources. A strategy for speeding things up is to break a job into pieces and then process the pieces simultaneously, in parallel.

In the six-input sorting network illustrated below (Activity 8 in the Computer Science Unplugged! book), students hold cards with the values 1 to 6 mixed up. They all move forward simultaneously, acting as parallel processors. The round circles are comparator nodes where two students meet and compare values. The person holding the smaller value goes left and the larger to the right. After everyone has traversed the network correctly (no one is left standing not knowing where to go) the numbers will be sorted in order.

In 2019, the world's largest human sorting network sorted the numbers from 1 to 50. There were over 1000 comparison nodes. Pupils from the International School Klosterneuburg and Sir Karl Popper School together with computer scientists from the Vienna Center for Logic and Algorithms (VCLA) of the Vienna University of Technology (TU Wien) set this world record. The activity was the prelude to project ADA which aims to bring computer science closer to young people and to foster gender mainstreaming in this discipline (https://www.vcla.at/2019/10/world-largest-sorting-network-based-on-cs-unplugged-world-record/).

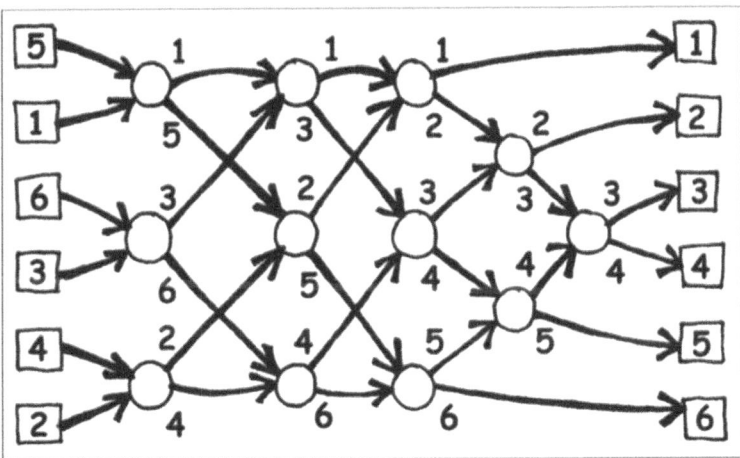

Fig. 3. Schematic of a 6-input Sorting Network

"Sorting algorithms are extremely important in information technology and appear in many programs and apps. Sorting networks are special forms of such algorithms that can also be implemented directly in hardware, for example on graphics processors," explains Prof. Stefan Szeider, initiator of the campaign from the VCLA of TU Wien to the Austrian Press Agency [11] (Fig. 3).

The sorting network can be used to sort anything that can be put in order: planets ordered by their distance to the sun, cities or countries ordered by size of population, musical notes, speeds, molecular weights of elements, heights of mountains, lexicographical or alphabetical order. The list is interdisciplinary.

Questions raised naturally by the students include the following. Does it matter that the smaller value goes left and the larger right? What if this is reversed? The sorting network in the picture can be seen to be symmetric, so for this 6-input sorting network, the answer is No. For 7 inputs, the best known sorting network is NOT symmetric, but the answer is still NO, because of the logical isomorphism of the meaning of "larger" and "smaller", a profound point (Fig. 4).

Fig. 4. A Sorting Network competition at the FPT Fest at the University of Bergen, Norway (Activity and photo thanks to Prof. Stefan Szeider, TU Wien).

When students reach the output where the values are in sorted order, they often want to mix up the values again and go back along the network. Does the network work in both directions? In the picture, imagine starting at the bottom with mixed up imports and following the arrows backwards. Little is known about efficient sorting networks that work in both directions of information flow.

A department chair in Multimedia and Design at the University of Newcastle, Australia used the sorting network to structure discussion to prioritize departmental objectives for the semester. Notice that not every two values meet each other on the 6-input network pictured. Not every professor met with every other professor. Can a different discussion system be designed usefully, perhaps where at each station three inputs meet and vote?

The sorting network has been used in Australian Aboriginal country with stories where the edges are trails on a walkabout and the comparator nodes are watering holes or billabongs [12]. Story and culture can be made part of the experience.

The picture in Fig. 5 shows the sorting network being used in front of the Frankfurt, Germany bourse in the banking district as a demonstration of the inequity between the homeless and the wealthy [13].

Fig. 5. The Sort Net in front of the Frankfurt Stock Exchange is being used to illustrate poverty and homelessness. (Activity and Photo by Playwright/Dramatic Artist Verena Specht-Ronique)

The iconic Sorting Network topic of CMSC math science dramatization transforms parallel computing algorithms into captivating walk-along pathways that are easy to follow and lead to open questions, unknowns about symmetry, permutations, and computational social choice, a major branch of modern mathematical science research with an annual international conference series.

The outreach activities of the authors using sorting networks, have led to exciting new research exploring 3-in / 3-out sorting networks as voting and priority-ordering social discussion systems. This is another example of how outreach is a two-way street that benefits core science as much as education and public communication.

2 Games that Cannot Go on Forever

One can use the mathematical concept of well-quasi-ordering, highly relevant to current research in algorithms and complexity, to create games that are easy to understand, addictive, and engage fundamental and powerful mathematical concepts. As we have found with Kid Krypto and other problems, the vigor of the two-way street continues. These games are about a superhot frontier of Parameterized Complexity (PC) with a key issue having to do with a class of games that are easy to play and to invent, and with rich opportunities for kid scientists.

In this section of the paper we describe the games and some of the open problems and give a taste of some of the kinds of research questions that are currently being grappled with by math scientists in the general area of PC and Artificial Intelligence (AI). We begin by giving a general description and the rules of play, and give several sample gameplays.

The following four rules define a game, $G(m)$.

1. The mathematical objects that constitute the moves of the game.
2. The order on the objects, a transitive relation $\leq$ on those mathematical objects.
3. The definition of the size of an object.
4. The required size of the first play object which we will denote m.

The *objects* are a set of mathematical entities, such as positive integers, graphs, or patterns of circles.

The *order* is a "less than or equal to" ($\leq$) operation that defines how to reduce a larger object to a smaller. The order is a transitive relation between the objects. If S represents the set of objects, then for all a, b, c $\in$ S, if a $\leq$ b and b $\leq$ c, then a $\leq$ c, the meaning of transitivity.

Players must agree on how to define the *size* of an object. For example, if the objects are graphs (a network of dots and lines), then the size might be the number of vertices, or it might be the number of edges, or it might be the number of vertices plus edges.

The *starting size*, which we will call 'm', is the size of the first move. This is agreed upon by the players.

There are two requirements for a legal next move in a gameplay of G. The games involve making moves by choosing (playing) an object of the required size, for that move. Each move must be of size one more than the previous move according to the definition of size. If the objects are graphs with size determined by the number of vertices in the graph, and the starting size m is chosen to be 3, then the first move will be of size 3 (a graph with three vertices). The second move will be of size 4 (a graph with four vertices).

The order, the transitive relation $\leq$, can often be conveniently defined in terms of a sequence of allowed operations that will take a larger object to a smaller one. For graphs, the operations might be deleting a vertex or deleting an edge. Each move must not be greater than or equal to *any* previous move according to the definition of the order. For example, the objects might be ordered pairs of positive integers and the order is the Cartesian product order. In the Cartesian product order, an ordered pair (a', b') is not greater than (a, b) if a' $\not\geq$ a AND b' $\not\geq$ b. The ordered pair (2, 6) is not larger than the pair (3, 4). The first term 2 is not larger than 3, so even though the second term 6 is larger than 4, the pair (2, 6) is not larger than the pair (3, 4) in the Cartesian product order because of the very important AND in the definition which means 'one or the other or both'.

To help understand the logic of the game, consider a game where the set of objects is the natural numbers, and the order (the relation between numbers) is $\leq$. Decide that the size of an object (a number) is the value of the number. Pick a number for the starting size of the first move, say 7. The size of the number 7 is 7, so 7 is the first move. The first rule of the game is that each move must be an object one size larger than the previous move. The only object that is of size one more than the number 7 is $7 + 1 = 8$, the number 8. So the size of the second move must be 8. The number 8 is forced to be the second move. But, this violates the second rule of the game that a legal move must not be greater than or equal to any previous move according to the order. Choosing 8 for the second move would violate this rule because $8 \geq 7$. Thus, the game has only one move, the first one of size 7. The length of the longest (and shortest) game is 1.

Most everyone is familiar with alphabetical ordering or the natural numbers with the usual *less than or equal to* ($\leq$) relation and where *every* pair of natural numbers is comparable (i.e., we know for the pair of numbers (5, 9) that 9 is greater than 5). Both the alphabetical ordering and the natural numbers ordering are *total linear orders*. A linear order is transitive: for any three natural numbers a, b, c, if $a \leq b$ and $b \leq c$, then $a \leq c$. A linear order is a special case of a *partial order*.

A partial order is transitive, but not all pairs of elements need to be comparable. There may be pairs (a, b) where neither $a \leq b$ nor $b \leq a$ holds. Consider the set of all subsets (the power set) of the set {1, 2, 3} with the subset relation ($\subseteq$). Not all pairs of subsets are comparable (e.g., {1} and {2} are not comparable because neither is a subset of the other).

A relation $\leq$ on a set S is a partial order if it is:

- Reflexive: For all $a \in S$, $a \leq a$.
- Antisymmetric: For all a, b $\in$ S, if $a \leq b$ and $b \leq a$, then $a = b$.
- Transitive: For all a, b, c $\in$ S, if $a \leq b$ and $b \leq c$, then $a \leq c$.

All of the games herein are based on a special sort of partial ordering known as *well quasi ordering* (wqo). The wqo order is reflexive and transitive, the elements are related by the $\leq$ relationship, and has a special additional property that every infinite sequence of elements has at least one pair of elements that *can* be compared according to the ordering. A set of objects that are wqo ordered will not contain an infinite set of elements that are mutually incomparable. Therefore, the list of moves of a gameplay for a wqo game G as defined by the four rules and the two rules of play cannot go on forever.

Even though the game must always end no matter how it is played, we may not be able to calculate when that will be. For some of these wqo games, we do not know the shortest possible length of a game with starting size m, or how long the game can go on before no one can make a legal move. The set S of objects with a less than or equal, transitive order on S that is a wqo sets up a Long Game function and a Short Game function. The functions are defined as the longest/shortest possible length (number of moves) of a game starting with a first move of an agreed upon size, m. The two game functions are important for very different reasons which will be discussed later, after we show some games.

Integer Sum Game with t Channels and First Move of Size m: ISG(t,m)

The first example uses 2 *channels* '(a, b)' of nonnegative integers. The game is defined with the following four rules.

1. Objects: 2 numbers represented as (a, b).
2. Order: defined in terms of operations, a move is not greater than or equal to any earlier move if it is not greater in both of the 2 channels, e.g., for two moves (a, b) and (a', b') it is not allowed for both $a \geq a'$ AND $b \geq b'$.
3. Size: the size of an object is the sum of the integers in the 2 channels.

4. Starting size: the size of the first move, which we denote by m.

In Fig. 6, the first move has two positive integers that add to a starting size $m = 6$. A different starting size could have been chosen, which would be a different game since the starting move object size m is one of the four rules defining the game. The order tells us how to decide if one move is greater than another. The order for this game is the Cartesian Product Order. The Cartesian Product Order is a partial order: it is reflexive, antisymmetric and transitive. The $\leq$ operation defined by this order transforms a larger object (a, b) to a smaller (a', b') if and only if $a' \leq a$ AND $b' \leq b$. The Cartesian product of the natural numbers N is denoted N x N.

The size of a move is the sum of the integers in the channels at any given play. At every move of the game it is important to check the two rules that make it legal.

For example, in the gameplay below (Fig. 6), the fifth move, made by Susie, satisfies the size rule. Her move meets the size requirement of size 10, since $1 + 9 = 10$, one size larger than the previous move of size 9. Secondly, it satisfies the order rule. Susie's move must not be greater than or equal to any previous move in the Cartesian product ordering of pairs of positive integers: the objects of play of the game. This is checked channel by channel for every previous move. The 1 in the first channel is smaller than any of the previous first channel moves (4, 3, 2, 8) so immediately we know that this is a legal move. We do not need to check the 9 in the second channel.

Integer Sum Game with Two Channels and Starting Size 6 (ISG (2, m))			
Move	**Size**	**Player**	**The Move**
First	6	Susie	(4, 2) The numbers sum to the required $m = 6$ for the size of the first move, by the definition of this game.
2nd	7	Timmy	(3, 4) The size is one more than the previous move size. The move is not $\geq$ any previous move.
3rd	8	Susie	(2, 6) A legal move. The sum is the required size 8 and the move is not $\geq$ any previous move.
4th	9	Timmy	(8, 1)
5th	10	Susie	(1, 9)

Fig. 6. It is not known how long this gameplay can continue, or even a strategy for finding out. But, the number of moves cannot go on forever! (Exercise: Is there a legal move of size 11? Is there a shorter gameplay if the first move is (5, 1) or some other pair that adds to $m = 6$?

Notice that it is very easy for humans to be led astray by confusing the logical statement "for every..., there exists ...." and the statement, "there exists..., for every". This is foundational logic for mathematics, law, and society generally.

We will denote the Integer Sum Two-Channel Game with starting size 5 by the notation ISG(2, 5) and understand that there is a 'family' of games where the number of channels could vary and the starting size is part of the definition of the game. By *gameplay*, we mean the record (history) of a particular series of legal moves for game G. The gameplay shown in Fig. 7 has a first move (0, 5) as defined by the game. The first move could be any of the alternatives: (5, 0) (1, 4), (4, 1), (2, 3), (3, 2), resulting in a different gameplay.

Three moves are shown and it is Timmy's turn to play a move of size 8. Timmy can check the list of the possibilities for the numbers in both channels that will sum to 8. Each possibility must be checked against every previous move. It must not be the case that the number in the first channel and in the second channel are greater than or equal to the number in the first channel and the second channel of any previous move. The move (5, 3) does not violate the first move because $3 \not\geq 0$. The move (5, 3) also does not violate the second move because $5 \not\geq 6$. The proposed move of (5, 3), which has the required size of 8, would not be a legal move because $(5, 3) \geq (4, 3)$ in the transitive order that defines the game, since $5 \geq 4$ AND $3 \geq 3$.

Integer Sum Game with Two Channels and Starting Size 5: ISG (2, 5)				(Exercise: Are any of these possibilities of a 4th move of size 8 a legal move?)
Move	Size	Player	The Move	
First	5	Susie	(0, 5) The starting size, 0 + 5 = 5.	(0, 8) (1, 7)
2nd	6	Timmy	(6, 0) The size goes up by one each move.	(2, 6) (3, 5)
3rd	7	Susie	(4, 3) A legal move. It is the required size 7 and is not $\geq$ than any previous move.	(4, 4) (5, 3) (6, 2) (7, 1)
4th	8	Timmy	(,)	(8, 0)

Fig. 7. This particular gameplay appears to be a short Integer Sum 2-Channel Game $m = 5$ with a gameplay with the starting move (0, 5). Another gameplay with starting move $m = 5$ could start with (1, 4), (2, 3), (3, 2), (4, 1) (5, 0). Would any of these starting moves result in a shorter or longer gameplay?

Figure 8 shows a game with three channels of positive integers. The game is defined with the same four rules: the objects, the order, the size and the starting size. The size of a move is the sum of the integers in the three channels at any given play. Again, the order for this game is the Cartesian Product Order, ' $\leq$ '. The Cartesian product of three natural numbers is denoted N x N x N.

1. The objects: a vector of 3 positive integers represented as (a, b, c).
2. The order: at every proposed move (a, b, c) of a particular gameplay, it is not allowed that $a \geq a'$ AND $b \geq b'$ AND $c \geq c'$ for any previous move (a', b', c').
3. The size of a move, a vector (a, b, c) of positive integers, is $(a + b + c)$.
4. The agreed upon size of the first move of the gameplay in Fig. 8 is $m = 7$.

For every proposed move of the gameplay, it is important to check the two criteria that make it a legal move. For the fourth move in the gameplay in the ISG(3, 7), we must check that Timmy's move is of the correct size. Yes, the values (1, 3, 6) add to 10.

Secondly, Timmy's move (his second move) must not be greater than or equal to any previous move. This is checked channel by channel for every previous move. The value 1 in the left channel of Tim's proposed move is ≥ 1 in both the first and second moves. The value 3 in the middle channel of Tim's proposed move is ≥ 1 in the middle channel of the first move and ≥ 3 in the middle channel of the third move. The 6 in Tim's right channel is greater the values in any previous move's right channel. Timmy's move (1, 3, 6) is not greater than the second or third moves, but it is not legal because in every channel it is $\geq$ to Susie's first move (1, 1, 3).

Integer Sum Game Three Channels with Starting Size 7: ISG(3, 7)			
Move	Size	Player	The Move
First	7	Susie	(1, 1, 5) The numbers $1 + 1 + 5 = 7$, the starting size m.
2nd	8	Timmy	(1, 5, 2) The size goes up by one each move, $1 + 5 + 2 = 8$.
3rd	9	Susie	(3, 3, 3)
4th	10	Timmy	(1, 3, 6) OOPS! NO GOOD, because the Cartesian Product of (1, 3, 6) is greater than or equal to (1, 1, 5) the first move, in *every* channel. Timmy tries again with (0, 0, 10). This is a valid move.
5th	11	Susie	(7, 3, 1)
6th	12	Timmy	(12, 0, 0)
7th	13	Susie	(1, 11, 1)

Fig. 8. This particular gameplay for the Integer Sum Game with Three Channels and Starting Size 7: ISG (3, 7) has only seven legal moves so far, from starting size $m = 7$ to size 13.

The Rooted Tree Game

An order on trees called 'upward topological containment' was proved by Joseph Kruskal and S. Tarkowski in 1960 to be a well quasi order. This leads to creating the Rooted Tree Game, defined with the upward topological containment order.

1. The objects are trees with a root at the bottom and trunk/branches going up.
2. The order is the topological containment order. Tree B is larger than tree A if tree A can fit inside Tree B by stretching, reflecting or moving the root upwards.
3. The size of a tree is the number of vertices in the tree.
4. The starting size is the agreed upon number of tree vertices in the first move.

Again, there are two rules that govern a legal move. The size (the number of vertices in the tree) of a legal move, has one more vertex than the number of vertices in the previous tree. And secondly, the move must not be $\geq$ any previous move (Fig. 9).

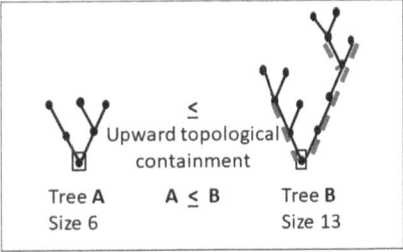

Fig. 9. Tree A fits inside Tree B stretching upward from the root on the right branch.

A graph can have several different drawings or pictures. The drawing of a graph is not the graph itself. A graph is an abstract mathematical entity with a relational structure of vertices and pairs of vertices called edges. Labeling vertices and noting which vertices they are connected to by an edge helps in identifying two different pictures of the same graph (Figs. 10, 11)

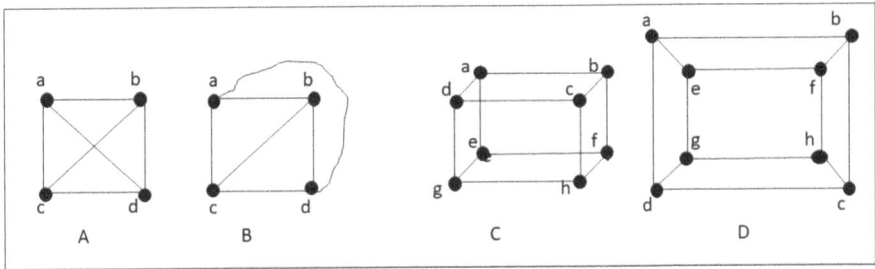

Fig. 10. A graph is the relationship between vertices and edges, not the drawing. Pictures A and B are drawings of the same graph. Pictures C and D are drawings of the same graph.

The Circle Game

Imagine that you are lying on the ground next to a tree and looking up along the trunk and into the branches of the tree. In the following figure, the circles on the right represent the tree on the left. The large circle is the root and the smallest circles are at the tree leaves.

Consider the circles as a rooted tree. There is a mapping (called an isomorphism or translation) between the tree and the circles such that the two structures or objects are fundamentally the same, even though they appear different in these two different ways of picturing the relevant relationship structure.

The proof of Kruskal's Theorem that rooted trees under upper topological containment are well quasi ordered by this mapping proves that the circles are well quasi ordered. The allowed operations that take a larger circles picture to a smaller circles picture (where by size we mean the number of circles) is circle deletion. At each move of the circle game we define, it is not allowed that circles can be erased to get an earlier move. A Circle Game cannot go on forever. That is what wqo is all about (Fig. 12).

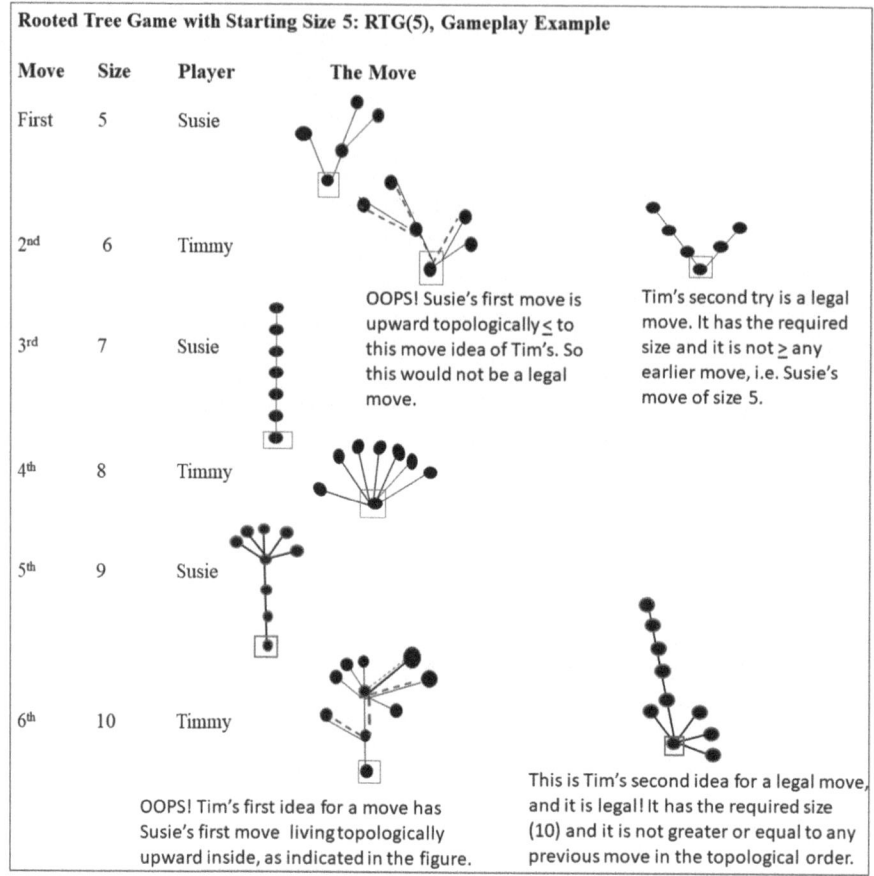

Fig. 11. A legal move must have the correct size. The move must be compared to every previous move. It must not be larger than any previous move (no earlier tree living inside).

The Circle Game has the following definition.

1. Objects: Circles pictures
2. Order: The operation of erasing circles will reduce a larger move to a smaller one. Circles may be concentric or side-by-side. They may not intersect. All that matters is the pattern of containments.
3. Size: The number of circles.
4. Starting Size: The number of circles in the first move circles picture.

A gameplay of the Circle Game with starting circle picture of size 4 is shown in the next figure. The gameplay has more than the five moves shown, but it is not known how many more before there is no next legal move. There are other moves possible for the second move. For example, Timmy could have drawn four concentric circles and one separate circle. Would this have shortened the gameplay? Would we get stuck sooner? (Fig. 13)

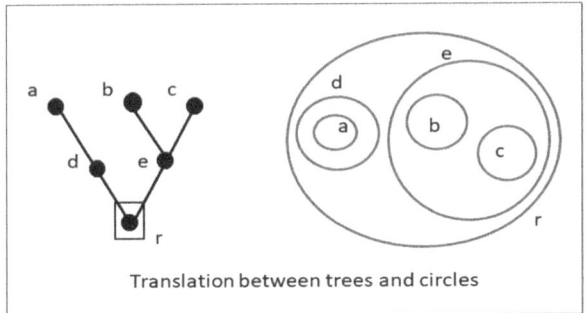

Translation between trees and circles

Fig. 12. In a translation from trees to circles, the root of the tree is the outermost circle. An embellishment of the game, not necessarily related to wqo, might be to add birds to the tree branches and to explore where and how to indicate this in a more elaborate circles picture representation (where the embellishment might be goats on mountains).

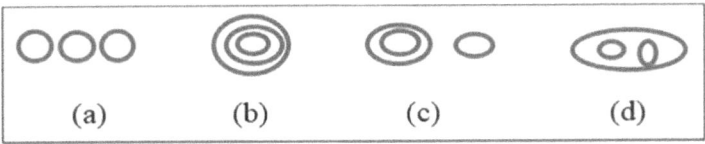

Fig. 13. There are four possibilities for a circles picture gameplay of size 3.

Seeing the translation between trees and circle pictures was an exciting recognition! It identified the wqo operation for the objects (circle pictures) of the Circle Game. Finding other such translations would be very interesting (Fig. 14).

A Cartesian Product Game

The Cartesian product of two sets A and B, denoted A X B, is the set of all ordered pairs (a, b) where a is in A and b is in B. For example, the Cartesian product of the four-element set {A, K, Q, J} with the four-element set {♠, ♥, ♦, ♣} returns a 16-element set {(A, ♠), (A, ♥), (A, ♦), (A, ♣), (K, ♠), ..., (Q, ♣), (J, ♠), (J, ♥), (J, ♦), (J, ♣)}. Why are Cartesian products (cross products) interesting and useful? Because the real world has more than one property.

The Cartesian product of well quasi ordered sets is well quasi ordered. This property allows us to create a Cartesian product game. The Cartesian product of games G_1 and G_2 (which we will denote G_1 X G_2) consists of legal move pairs (M_1, M_2) where.

- M_1 is a legal move in G_1
- M_2 is a legal move in G_2, and the size of the move is the sum of the sizes.

The objects in the Integer Sum Game with Two Channels are pairs of natural numbers. This could be thought of as a Cartesian product game with Game A having objects the natural numbers and Game B having objects the natural numbers. The Cartesian product of the natural numbers is represented as ℕ X ℕ.

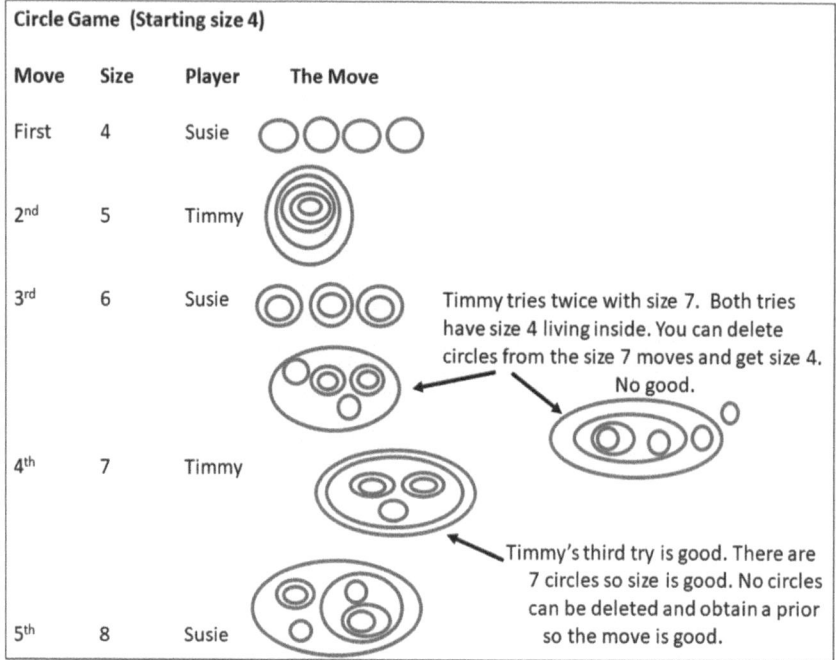

Fig. 14. At each move, it is not allowed that circles can be erased to obtain an earlier move.

We can have a well-defined game once we have decided on the four rules. The following four rules define a game made by taking the Cartesian product of the Circle Game and the Integer Sum 2-Channel Game.

1. The objects are pairs (L, R), where L is a circles picture and R is a pair of non-negative integers.
2. The order for the circles is circle deletion, and for the integers it is $\leq$ as before.
3. The size of a move is the sum of the sizes from the two games: $M_1 + M_2$, i.e., the number of circles from the Circles Game (the left factor L) + the integer sum from both channels of the Integer Sum 2-Channel Game (the right factor R).
4. The starting move size m.

We check if the second gameplay by Timmy in Fig. 15 is correct. Timmy's move has the correct size of five, the number of circles in the circles picture half of the move (the Left part of the move) plus the size of the Right part of the move (which has size (0 + 0) = 0). Tim's move has size (L game + R game) which is $5 + 0 = 5$, the required size.

In the left factor of Timmy's move, the five circles are $\geq$ than the one circle in the first move so this might mean this is not a legal gameplay. Four of Tim's circles could be erased and his left factor would be equal to Susie's left factor, or all of Tim's left factor circles could be erased and then his left factor would equal the left factor of the moves of size 3 and of size 4.

Move	Size	Player	The Move
First	2	Susie	[○ , (1, 0)] The one circle from the Circles picture and the 1 from the Integer Sum Game add to 2, the starting size
2nd	3	Timmy	[∅ , (1, 2)] The empty set in the Circles picture indicates that there is no circle, a value of 0. The move has the required sum: 0 + 3. The empty set cannot be reduced to the circle in the first move, so this is a legal gameplay.
3rd	4	Susie	[∅ , (0, 4)]
4th	5	Timmy	[○○○○○ , (0 , 0)] Susie has proposed this move for size 6. Is it legal?
5th	6	Susie	[○○○○ , (2, 0)]

Circle Game X Integer Sum 2-Channel Game (Starting size 2)

Fig. 15. Cartesian Product of a Circle Game with an Integer Sum 2-Channel Game with Starting Size 2. (Exercise: verify that the moves 2, 3, 4, 5 are legal. Is there a legal move of size 6?)

However, in the right factor of Tim's move, the zero in the first channel of the Integer Sum is not greater that the '1' in the first channel in the first or second move. Hoorah! Tim's move is not ≥ the first or second move. It is legal. Furthermore, it is not greater than the 3rd move because 0 is not > 4 in the right channel of the 3rd move's right factor.

Susie's proposed 3rd move has the correct size 6. However, it is ≥ in every channel of the first move, and therefore the move is not legal. Is there a legal move of size 6?

We do not know at what point we will not be able to make another legal move in the above Cartesian Product gameplay. No matter how cleverly we play, if we continue this gameplay of the game, it will become impossible to make another legal move. We cannot make the game go on forever!

3 A Few Examples of What Little is Known About WQO Short and Long Game Functions.

For every WQO game G, as specified by its four defining rules (including the required size m of the first move) there are two natural and extremely interesting associated functions:

- $G_S(m)$, the short game function, and
- $G_L(m)$, the long game function.

Little is known about these functions for most natural well quasi orders, but they are extremely addictive open problem frontiers, accessible to kids (and others, including ourselves!). To play one of these games requires acute attention to (alternating) logical quantification.

We know that WQO games cannot go on forever no matter how the game is played. One can prove that for any WQO game G (with starting size m) and just attending to the first three rules and ignoring m for the moment, there exists an absolute barrier bound M, a positive integer, for which any gameplay for game G with starting size m, cannot go on beyond.

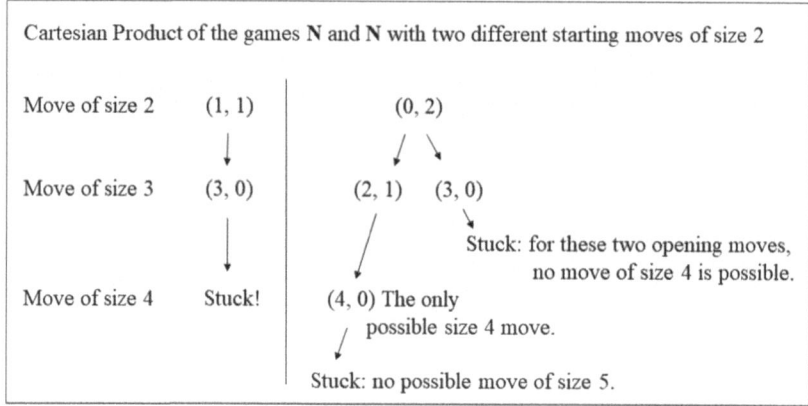

Fig. 16. The only possible move of size 3 following (1, 1) is (3, 0). There are two possible moves of size 3 following (0, 2). (Exercise: Argue this! Argue that there is no move of size 5).

For a specific game legislated by the four rules and with a starting size m, how short or long a game could we play before getting stuck and not able to make another legal move? The two gameplays of the 2-channel Cartesian Product game with starting size 2 shown in Fig. 16 seem to be the only ones possible starting from a first move of size 2, giving a short game of length 2 and a long game of length 3.

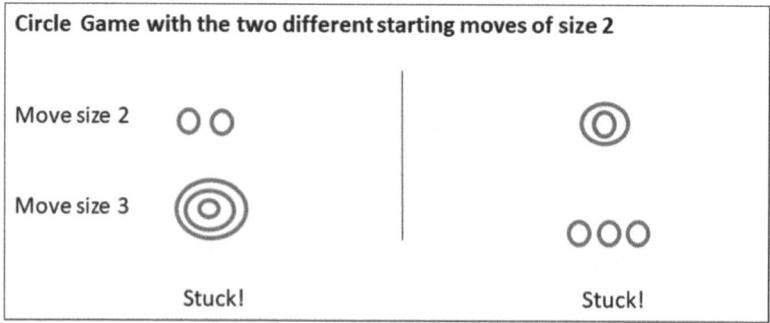

Fig. 17. Both possibilities of an opening move of size 2 lead to a gameplay that becomes stuck after two moves. (Exercise: argue this! No move of size 4 is possible. The short game function is equal to the long game function.

The Circle Game with starting size 2 has only two possibilities for a first move gameplay: two circles side-by-side or two concentric circles. For the second move of size 3, the gameplay of two circles side-by-side is forced to be followed by a move of three concentric circles. Then, stuck! The other possibility, a first move of two concentric circles, is forced to be followed by three circles side-by-side. Then, stuck! The short game function for the Circle Game with starting size 2 is equal to the long game function. Games that have the short game function equal to the long game function might be quite unusual, but we don't know.

The next figure shows a gameplay for the Cartesian product of the Circle Game and the Integer Sum 2-Channel Game with starting size 2. The gameplay starts with a first move of two circles on the left, and zero (from the Integer Sum game) on the right. A possible way to describe the distribution of the move is to denote the Circle Game sum on the left and the nonnegative Integer Sum Game sum on the right as [L, R] or [2, 0]. The distribution of the second move (of size 3) would be denoted [3, 0], the distribution of the move of size four would be [0, 4] and the distribution for the move of size five is [2, 3].

Future gameplay moves of a game depend on the history (the gameplay) of the moves that have already been played. Can we find a strategy to predict size distributions that give legal move sums and will lead to a short game? This is a brain-rich question for kids.

We know short gameplays for the Integer Sum Game with starting size 2 (Fig. 16). Also, we know short gameplays for the Circle Game with starting size 2 (Fig. 17). Does this knowledge help us to find a short game for the Cartesian Product, such as the gameplay in Fig. 18?

Does knowing a short game for Game A and a short game for Game B, both with starting size m, give us any clues on finding moves that will shut down any gameplay quickly, a short game with starting size m formed by their Cartesian product, Game A x Game B with starting size m, for the product game?

Circle Game X Integer Sum 2-Channel Game Starting size 2			
Move	Size	Player	The Move
First	2	Susie	 [○ ○ (0, 0)] The number of circles is 2, the starting size m.
2nd	3	Timmy	 [◎ , (0, 0)]
3rd	4	Susie	[∅, (0, 4)] The empty set plus zero plus four equals 4, the required size. The empty set is not greater than or equal to either of the Left factors of any previous move.
4th	5	Timmy	 [◎ , (1, 2)]

Fig. 18. Does knowing a short gameplay for the Circle Game with starting size 2 and a short gameplay for the 2-channel Integer Sum Game with starting size 2 help in finding a short gameplay for the Cartesian Product game with starting size 2? (This game has a different starting move than that in Fig. 15. It is a different gameplay.)

Branching or looking at all the cases to find all possible moves of a short game, i.e., finding all possible size ten moves following a previous move of size 9 might be possible, but for a long game this method would soon blow away all computer clicks imaginable even on a quantum computer.

The long game function and the short game function are both important but for different reasons. Interest in the long game function has to do with the power of axiom systems. Mathematician Harvey Friedman defined these ideas about the long games in terms of fast growing functions. The asymptotic growth rate of super fast-growing functions (the long games) grows so fast that, depending on your axiom system, you will not be able to prove anything about them. Every axiom system has a speed limit and these functions for some games (very non-trivial to prove) grow beyond the speed limit for the axiom system of Peano arithmetic, the axioms that define the arithmetical properties of natural numbers.

Interest in the short game function has to do with how fast practical combinatorial convergence might happen for FPT algorithms based on the obstruction set learning algorithm of [15]. (See also [16]).

Investigating the short game and long game functions (as a function of the first move size m) is a rich area of mathematical unknowns accessible to investigation by kids, and of solid interest in current research.

4 Conclusion

The primary purpose of this paper has been to describe a class of mathematical games based on the well quasi ordering of various mathematical objects. To play one of these games, one must seriously engage with logical quantification and alternating quantifiers (*for every* and *there exist*), which is central to mathematical thinking (famously and prominently in the foundations of calculus, and analysis generally).

The games are entertaining and addictive and they can be further developed in a wide variety of ways. We believe they have significant commercial potential.

We have tried them out with children we met while traveling on trains in Europe, children as young as age four who barely knew how to draw circles became engaged with the game understood the logic.

Besides posing fun intellectual challenges in playing them, the games pose fascinating unexplored mathematical frontiers – the concrete challenges of finding bounds on the length of short and long games. Short game bounds are directly related to combinatorial convergence issues for encountered – obstruction learning algorithms for FPT algorithms based on well-quasi-ordering [14–16], a hot area of current research in algorithms and complexity, deeply related to mathematical methods in Artificial Intelligence.

A version of this paper with additional details is posted on arXiv.

References

1. Casey, N., Fellows, M.R.: This is Mega-Mathematics! Los Alamos National Labs (1992). http://www.ccs3.lanl.gov/mega-math/write.html

2. Fellows, M.: Research on mega-math: Discrete mathematics and computer science for children, final report to the Los Alamos National Labs (1995). https://digital.library.unt.edu/ark:/67531/metadc621790/m1/3/
3. Bell, T., Witten, I., Fellows, M.: Computer Science Unplugged: Off-line activities and games for all ages (original book) (1999), http://csunplugged.org. The "Classic" (original) book can be accessed a couple of ways: Go to https://www.csunplugged.org/en/, click on "Topics", and the last topic is "Classic". Then click on "Books". Go directly to https://classic.csunplugged.org/books/ (download the 2015 edition).Get the PDF directly at https://classic.csunplugged.org/documents/books/english/CSUnplugged_OS_2015_v3.1.pdf/
4. https://www.claymath.org/millennium-problems/. Accessed 12 Aug 2024
5. https://www.washingtonpost.com/entertainment/movies/2023/10/31/radical-movie-review/. Accessed 12 Aug 2024
6. Rosamond, F. A.: 7th International Conference on Creative Mathematical Sciences Communication (CMSC'24), Journal of Humanistic Mathematics, vol. 14 Issue 2, pp. 683–686 (2024). https://doi.org/10.5642/jhummath.EXDH9190, https://scholarship.claremont.edu/jhm/vol14/iss2/29
7. Bell, T., Thimbleby, H., Fellows, M., Witten, I., Koblitz, N., Powell, M.: Explaining cryptographic systems to the general public. Comput. Educ. **40**, 199–215 (2003)
8. Rosamond, F. A.: Computational thinking enrichment: public-key cryptography. Inform. Educ. **17**(1), 93–103. Vilnius University (2018)
9. Pre-University Math Education: Cryptography As a Teaching Tool, bottom of Neil Koblitz' website, https://sites.math.washington.edu//~koblitz/. Accessed 12 Aug 2024
10. Fellows, M., Koblitz, N.: Combinatorial cryptosystems galore! Contemp. Math. **168**, 51–51 (1994)
11. Website https://www.ada.wien/aktivitat-8/. Accessed 12 Aug 2024
12. Rosamond, F.A.: The Australasian Council of Deans of Information and Communications Technology (ACDICT), Creative Mathematical Sciences Communication Indigenous Community, Report (2023)
13. Website www.csmaths.org. Accessed 12 Aug 2024
14. Fellows, M., Langston, M.A.: Nonconstructive tools for proving polynomial-time decidability. JACM **35**(3), 727–739 (1988)
15. Fellows, M., Langston, M. A.: On Search, Decision and the Efficiency of Polynomial-Time Algorithms (Extended Abstract) Proceedings of the 21st Annual ACM Symposium on Theory of Computing, 14–17 May 1989, Seattle, Washington, USA (1989)
16. Fellows, M., Rosamond, F.A.: Collaborating with Hans: some remaining wonderments. In: Fomin, F.V., Kratsch, S., van Leeuwen, E.J. (eds.) Treewidth, Kernels, and Algorithms. LNCS, vol. 12160, pp. 7–17 (2020). https://doi.org/10.1007/978-3-030-42071-0_2

# Tactile Learning: Unplugged Graphs, Trees, and Patterns

Unplugging Dijkstra's Algorithm as a Mechanical Device

Riko Jacob[1] and Francesco Silvestri[2](✉)

[1] IT University of Copenhagen, Copenhagen, Denmark
rikj@itu.dk
[2] University of Padua, Padua, Italy
silvestri@dei.unipd.it

Abstract. Graphs are a fundamental concept in computer science, effectively modeling diverse scenarios such as social networks, protein interactions, and mobility. Dijkstra's algorithm is crucial for computing single-source shortest path in graphs and is a key component of graph processing. This paper presents an educational activity designed to "unplug" graphs and Dijkstra's algorithm, making these topics accessible to a broad audience. The activity utilizes a physical graph with chains as edges and key rings with retractable badge holders as nodes. By pulling two nodes of this graph apart, it is possible to find a shortest path between these nodes. This can be used to visualize how Dijkstra's algorithm works, including how the graph models the world. It invites for discussing how much more efficient this is compared to enumerating all paths, and what additional insights computer scientists had to achieve impressive speedups over plain Dijkstra, allowing for route planning to be perceived as a solved problem, where we use the packaged solution without further thought. We discuss the implementation of this activity in public outreach events, such as Culture Nights and primary school classrooms.

Keywords: Dijkstra's algorithm · mechanical device · CS Unplugged

1 Introduction

Computer Science Unplugged (CS Unplugged) is a collection of teaching activities designed to introduce key concepts from Computer Science without using digital devices or programming, but instead exploiting simple materials (pencils, paper, cards, ...), games, and kinaesthetic engagement. CS Unplugged was introduced in the seminal work [4] by T. Bell, I. H. Witten, and M. Fellows which presents a variety of activities aimed at introducing computer science ideas to primary school students. Since then, hundreds of works have proposed activities covering a wide range of topics, including binary encoding, sorting, image representation, cryptography, and more. Many of these activities are accessible online through websites, research papers, or simple documents (see, e.g., www.csunplugged.org, [1,2] and references therein). Although initially intended for primary school students, the activities have found broad applications in public outreach efforts.

The concept of graphs is highly significant in computer science, as it can model various scenarios, including applications in social networks, bioinformatics, and mobility. However, a non-technical audience, in particular primary school kids, might incur difficulties in understanding graphs and how they can model reality. To address this challenge, several CS Unplugged activities in the literature aim at introducing the concept of graph, some graph primitives like Minimum Spanning Tree, and intractable problems like graph coloring or dominating set. These activities are mostly based on paper and pencil activities, which require solving challenges like coloring a map or positioning ice cream vans on a map. Dijkstra's algorithm is among ones of the most studied graph processing algorithms and it solves the single-source shortest path problem: given a graph $G = (V, E)$ and a source vertex $s \in V$, find all shortest paths from s to every vertex $t \in V$. There are numerous educational videos and web pages aimed at explaining the intuition behind Dijkstra's algorithm and how to implement it: however, to the best of our knowledge, there is no previous work aiming at unplugging Dijkstra's algorithm and introducing it without any digital devices or coding.

In this paper, we propose an unplugged activity for introducing the concept of graph and Dijkstra's algorithm using a mechanical device; the activity has already been used in several events, but never formally described. The novelty of the approach is the use of a physical graph constructed on a board, with chains as edges and rings with retractable badge holders as vertexes (see Fig. 1). A map is plotted on the board's surface, representing a geographical region of interest to the audience: the vertexes represent points of interest (e.g., cities), while the edges capture connections on the map (e.g. roads or train lines). The device allows one to find the shortest paths by simply vertically pulling up vertexes. To find the shortest path between two vertexes s and t of the graph, it suffices to take the two vertexes and pull them apart: the shortest path between s and t is going to be the taut chains that are suspended over the other chains, i.e., the path of chains that stops the two vertexes from being able to be pulled further apart. Moreover, by just pulling up one vertex s, all other vertexes are raised (basically) in order of increasing distance from s, and the shortest path from s to a vertex t can be easily detected as it is the only taut chain path between the two nodes. When pulling a vertex, we are physically implementing Dijkstra's algorithm using the mechanical device, providing a visualization of the algorithm.

This activity can build on common knowledge like navigator apps and maps, and it can be used to introduce several concepts. Indeed, the device provides an example of how the concept of graphs captures real-world problems, like finding the best train combination to move from one city to another. Moreover, the device facilitates discussions on the complexity of solutions based on an exhaustive search for route planning, and on the significant improvements provided by fast solutions such as Dijkstra's algorithm. The goal of this paper is to describe the device and the associated activity, and how the device was constructed and used in different scenarios. The paper is structured as follows: in Sect. 2, we introduce some graph terminology used in the paper and relevant previous work; then, in Sect. 3, we describe the activity, the relation to Dijkstra's algorithm, and how to construct the device; in Sect. 4, we describe our experience of the table in public outreach public in Culture Nights, and in fifth-grade classes (about 10–11 years) in an Italian Montessori primary school; we conclude with some final remarks in Sect. 5.

2 Preliminaries

2.1 Graphs and Dijkstra's Algorithm

In this section, we introduce some graph notations and Dijkstra's algorithm; note that these concepts are not required for the main activity, which can be presented without any technical requirement. A *graph* is a pair $G = (V, E)$, where $V = \{v_1, \ldots v_n\}$ is a set of elements called vertexes and $E \subseteq V \times V$ is a set of m unordered pairs from V, whose elements are called edges. A *weighted graph* is a graph where each edge $e \in E$ is given a weight $w_e \in \mathbb{R}$; we assume weights to be non-negative. A *path* $\mathcal{P}$ is a sequence $(v_{i_1}, v_{i_2}, \ldots v_{i_\ell})$ of vertices in V such that there exists in E an edge $(v_{i_j}, v_{i_{j+1}})$ for each $j = 1, \ldots, \ell - 1$. The *weight* (or cost) of a path $\mathcal{P} = (v_{i_1}, v_{i_2}, \ldots v_{i_\ell})$ is the sum of the weights of the edge among the path, that is $\sum_{j=1}^{\ell-1} w_{(v_{i_j}, v_{i_{j+1}})}$. Given two vertices $s, t \in V$, the *shortest path* from s to t is a path with minimum weight among all paths from s to t; we denote with $d(s,t)$ the weight of this shortest path.

Given a graph $G = (V, E)$ and a vertex $s \in V$, the *single-source shortest path* (SSSP) problem requires computing the shortest path from $s \in V$ to every vertex $t \in V$. The *Dijkstra's algorithm* is one of the most common solutions to SSSP, and it was conceived by E. W. Dijkstra in 1956 [6]. We provide a simple explanation of the algorithm and we refer to [5] for a more detailed analysis. The algorithm iteratively constructs with a greedy approach a set $S \subseteq V$ of vertices for which we know a shortest path from the source s; for all vertexes $t \in V \setminus S$, we instead have an upper bound $\tilde{d}(s,t)$ on the cost to reach t from s by using only vertices in S; the bounds are improved in each iteration until we reach the exact values.

More specifically, we initially have $S = \{s\}$ since the source is connected to itself, and the upper bound $\tilde{d}(s,t)$ is set as:

$$\tilde{d}(s,t) = \begin{cases} w_{(s,t)} & \text{if } (s,t) \in E \\ +\infty & \text{otherwise} \end{cases}$$

Then, the algorithm repeats $n-1$ times the following steps:

1. Add to S the vertex u in $V \setminus S$ with minimum upper bound on the cost; for this node, we indeed have $d(s,u) = \tilde{d}(s,u)$.

2. For each edge (u,t) with $t \in V \setminus S$, update the upper bound as follows:

$$\tilde{d}(s,t) = \min\{\tilde{d}(s,t), d(s,u) + w_{(u,t)}\}$$

We observe that the above algorithm exploits the fact that when u is added to S, then the current upper bound $\tilde{d}(s,u)$ is the cost of the shortest path $d(s,u)$. Although the above algorithm only computes the cost of the shortest path, it can easily be restated to compute the actual shortest path.

2.2 Previous Work

There are numerous CS Unplugged activities documented in various websites and academic papers, as well as many activities that have yet to be formally described. For a more comprehensive examination, we recommend referring to the CS Unplugged website and book [4], as well as several surveys (e.g. [1–3]). For an overview of the different applications of CS Unplugged and discussions on the effectiveness and limits of the unplugged approach, we refer to the aforementioned [3] and reference therein. Given the importance of graphs and graph primitives in computer science, several unplugged activities on these topics have been presented. Already Bell, Witten, and Fellows in the book *Computer Science Unplugged* [4] provide several activities on graph primitives, such as Minimum Spanning Tree (MST), sorting network, graph coloring, dominating sets, Steiner trees.

In the MST activity, students are given a map of houses connected by muddy roads and are asked students to pave enough streets so that it is possible for everyone to travel from their house to anyone else's house only along paved roads; clearly, the paving cost should be as little as possible. The map with houses and roads nicely introduces the notion of graphs to kids. Moreover, the paving problem is an intuitive example of MST, that can be easily solved on small graphs, but also on slightly larger with Kruskal's algorithm. Sorting networks are an example of direct acyclic graphs used for describing computations: the activity asks students to sort elements by following the paths given by a sorting network drawn on the floor. The main goal of this activity is not to introduce graphs, but rather to reason about sorting algorithms; in fact, the graph structure of paths is usually not emphasized.

While the above activities focus on algorithm design, the activities on graph coloring, dominating sets, and Steiner trees focus more on intractability by showing that simple problems might be complex to solve even with small inputs. As in the previous activity, the activities build on a particular problem (e.g., setting ice cream vans, or coloring a map) to introduce the concept of a graph. Then, different input examples of increasing size can be given to the students: a warm-up with easy input instances allows students to better understand the problems; then, larger input instances challenge the students and show that finding a solution quickly becomes harder as soon as the input size increases.

Most activities on graphs are based on a "paper and pen" approach. An exception is provided on activities on sorting networks that can be seen as a kinaesthetic learning activity. To the best of our knowledge, there is no previous work on describing graph primitives, like Dijkstra's algorithm, using a mechanical device as proposed in this paper.

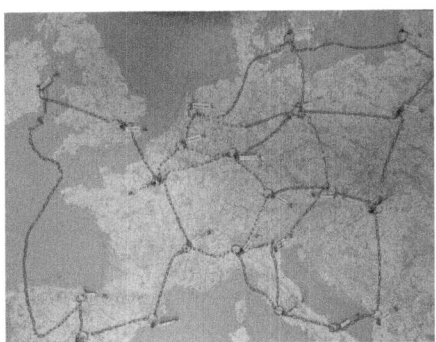

Fig. 1. The board used for explaining Dijkstra's algorithm: the left picture shows the complete graph on the board and the right picture how it was used during an activity.

3 Activity Description

In this section, we describe the activity to introduce graphs and Dijkstra's algorithm with the mechanical device. We first describe how the activity was presented to two fifth-grade classes of an Italian primary school (about 10–11 year-old students). Then, we explain why the device implements Dijkstra's algorithm. Finally, we describe how to construct the mechanical device.

3.1 Narrative

Graphs and Paths. The activity leverages on the board in Fig. 1, which represents (part of) Europe. At the top, a set of rings highlights some cities: each ring has the name of the city and it is connected to the table via a retractable badge holder. Each holder allows to pull up the ring with the city name from the board, and then easily reposition it on the board; this feature will be useful later. The chains represent the *trains* or *ferries* that connect two cities (in both directions). The length of the chain (i.e., the number of links) represents the travel time from one city to another using that train/ferry: for instance, the chain representing the train from Milan to Lyon is much shorter than the ferry from Bilbao to Dublin, meaning that the ferry takes much more time than the train; we can say that each link is about 30 min. The rings and chains create a *graph*: the rings (i.e., cities) are called *vertexes*, the chains are named *edges*, and the traveling time is the *weight* of the edge. A sequence of trains/ferries for moving from one city to another one is called *path*: for instance, we can travel from Madrid to Brussels by taking the trains Madrid-Paris and Paris-Brussels. The total traveling time of the path is proportional to the number of links in the chains we use: the larger the number of links in the path, the longer the traveling time between the two vertexes. We assume that we do not spend time on a vertex for a train transfer.[1]

Shortest Paths. Let's say we start from Padua and we want to reach Munich: which trains should we take if we want to minimize the traveling time? There are many train combinations: we can go via the paths Padua-Rome-Milan-Munich or Padua-Milan-Lyon-Paris-Frankfurt-Munich, which however do not minimize the traveling time. A quick look at the board shows that the best paths seem to be Padua-Milan-Munich or Padua-Wien-Munich. As the traveling time is given by the length of the chains, by counting chain links, we easily see that the fastest solution from Padua to Munich is via Milan (21 vs 41 chain links). A path that minimizes the total traveling time between two cities is named *shortest path*. We observe that for providing the exact solution, we have considered several train combinations: some combinations clearly take longer times, but others require a more fine analysis by counting chain links. If we now take two far away cities, like Padua and Copenhagen, we see that the set of possible train/ferry combinations is very large and we need a long time to find the shortest path by enumerating all possible train/ferry combinations.

Fig. 2. Two examples of activities on the mechanical device. On the left, two cities are pulled up at the same time and the top path denotes the shortest path between the two cities. On the right, we pull up one vertex: here, we are mechanically implementing Dijkstra's algorithm.

Shortest Path Between Two Vertexes. To easily determine the shortest path, we can use a simple mechanical approach. Suppose we want to travel from

---

[1] Although the map is inspired by some existing train/ferry lines, the graph does not represent an existing network nor the traveling times between cities are real.

Paris to Warsaw: we vertically lift the rings of the two cities, and the path that rises above the other chains is the shortest one. More specifically, take the two rings associated with the two cities and slowly lift them. As we do, the chains start to rise and, at some point, a path connecting the two cities appears with its chains taut and above the others: this path is the shortest one. Refer to the left image of Fig. 2 for an example.

At this point, we can challenge the participants by asking them to determine the shortest path between different cities or after changing the chain lengths. For instance, which is the shortest path between Milan and Frankfurt if there is a long delay on the Munich-Frankfurt train (i.e., we replace the chain with a longer one)? If the tunnel between Paris and London closes for maintenance, what is the shortest path from Madrid to London? (In this case, the only available path is represented by the very long chain of the ferry from Bilbao to Dublin).

Single-Source Shortest Path and Dijkstra's Algorithm. To better understand why this approach yields the shortest path, let's assume we are in Frankfurt and want to compute the shortest path from Frankfurt to *every* city in the graph. This problem is called the *single-source shortest path*. If we slowly lift Frankfurt's ring, we see some chains rising, and eventually, another city's ring lifts. This city represents the closest one to Frankfurt in terms of travel time (Munich, in our case). By pulling up Frankfurt's ring, we lift its edges, and the shortest edge is the first to rise completely. Additionally, this edge is the only one fully stretched, while all other edges remain slightly loose. Refer to the right image of Fig. 2 for an illustration.

If we continue lifting Frankfurt's ring, another city rises, specifically the second closest city to Frankfurt (Brussels, in our case). The shortest path from Frankfurt to this second city is given by the only taut path between the two cities. We observe that there cannot be any shorter path from Frankfurt to the second city; otherwise, the pulling would have already stretched this path.

If we continue, we find the shortest paths from Frankfurt to all the other cities. Additionally, by observing the order in which the cities rise, we see them sorted by travel time: the shortest path from Frankfurt to another city is given by the only taut path between the two cities. What is happening here? By pulling one vertex, we are mechanically implementing Dijkstra's algorithm, one of the most important algorithms in computer science. This algorithm computes the shortest paths from one vertex, called the source, to all other vertices in the graph.

Consider now the previous case where we want to compute the shortest path from Paris to Warsaw. We can simply lift Paris's ring: when Warsaw's ring rises too, we have found the shortest path between the two cities. We could achieve the same result by lifting Warsaw's ring instead of Paris's. When we pull both rings, we are applying Dijkstra's algorithm from both cities simultaneously. At some point, a rising path starting from Paris and a rising path starting from Warsaw meet (approximately in the central part of the shortest path), creating a single suspended path between the two cities. This approach allows us to better detect the shortest path, as it is completely hanging. Additionally, it requires less lifting compared to using a single ring and enables us to compute the shortest path even when the string in the badge handler is not sufficiently long.

3.2 What's Behind?

We now explain why the above mechanical device is implementing Dijkstra's algorithm. As recalled in Sect. 2.1, Dijkstra's algorithm iteratively computes a set of vertexes S which contains all vertexes of the input graph $G = (V, E)$ for which we already know the shortest path from the source $s \in V$. When we slowly pull up the source vertex (e.g., Frankfurt), we are iteratively constructing S: specifically, all hanging vertexes are the ones in set S. In the beginning, only the source s is up, like in the initial step of Dijkstra's algorithm: the source is connected to itself by a path of length 0. By continuing lifting s, we apply a pulse to the chains that propagate with uniform speed among all edges adjacent to the source s: therefore the second node to rise is the one connected to s by the shortest edge; this is the second vertex added to S.

Assume that we have already raised $|S| \geq 2$ vertexes and we continue to lift s. The pulse propagates among all edges that connect a node in S with a node in $V \setminus S$. Let v be the next vertex to rise; we claim that v is the vertex with the smallest shortest path from s among the vertexes in $V \setminus S$. If this was not true, there would exist another vertex $w \in V \setminus S$ that lies on the board when v raises and with $d(s, w) < d(s, v)$. However since the pulse propagates with uniform speed, we must have that w would have raised before v, which is a contradiction. As Dijkstra's algorithm adds the vertex in $V \setminus S$ with the smallest shortest path, then v is also the next vertex added to S.

3.3 Construction of the Device

Now, let's explain how the mechanical device is constructed. It consists of a particle board: as a reference, the board in Fig. 1 measures 120 cm by 90 cm in size. Due to its weight, it may be cumbersome to move: to address this, we suggest adding a handle on the back for easier carrying, or dividing the board into two folding parts connected by hinges. The board is supported by two wooden folding sawhorses.

At the top of the board, we affix a map printed on an adhesive vinyl sheet. The specific geographic area depicted on the map can be chosen based on the audience and the educational context, as well as the dimension of the board. For example, at a Culture Night event in Copenhagen, we used a map of the city: this choice was ideal because the audience was familiar with local street names and with the opening/closing of bridges that affect traffic flows. Conversely, when working with fifth-grade students at an Italian Montessori school, we used a map of Europe: this decision aimed to complement their recent introduction to European geography and to spark curiosity about cities in different countries. Maps for the board can be obtained from OpenStreetMap using tools such as Big

Map 2 (https://github.com/zverik/bigmap2), which allows downloading high-resolution maps suitable for large displays.

Each vertex on the top of the board is represented by a key ring connected to a retractable badge holder, which is positioned on the bottom part of the board. We repeat these steps for each vertex:

1. A small hole is drilled on the board where to position the vertex.
2. The retractable badge holder is placed on the bottom part of the board.
3. The thread from the badge holder is passed through the hole drilled on the board.
4. The thread is then knotted onto a ring on the top part of the board.
5. A label with the city name is added to identify the vertex.

The tension of the retractable holder keeps the ring attached to the board, but it still allows to raise it. It's important to note that during this operation, the end of the badge holder where badges are normally attached needs to be cut. Care must be taken during this step to prevent the thread from irreversibly retracting inside the holder once the end is cut.

Edges in our setup consist of metal chains with small links (e.g., 18 mm). An alternative option is using non-elastic ropes. Chains are chosen for their ease of attachment to rings and the convenience of measuring length by counting links. To ensure simplicity, the graph should be planar to avoid complications where crossing edges might interfere with the pulling operation. Finally, we remark that the diameter of the rings might affect the pulling operation (i.e., by pulling a path slightly longer than the shortest path): however, as soon as the rings are significantly smaller than chains, the effect is negligible.

4 Case Studies

In this section, we describe our experience in using the mechanical device for public outreach. We remark that this is anecdotal evidence, and should not be seen as an experimental evaluation of the proposed method.

4.1 Broad Audience

The first version of the described device was designed and built for an exhibition for the 25th anniversary of the Computer Science Department at ETH Zurich, and subsequent versions for Culture Nights at the Technical University of Munich and the IT University of Copenhagen. For the exhibition, we additionally had rice visualizing the number of top-to-bottom paths in a Manhattan-like graph of side-length n. We have many good experiences showing that the device is a nice activity for the general public. Given the size of the device, it attracts some attention. The explanation can easily be adapted to whatever pre-knowledge visitors come with. The robust mechanical design means that even young children will not damage it.

4.2 Primary School

The board was used within a 1.5-hour activity for two fifth-grade classes of the *Scuola Primaria M. Montessori* in Padua, an Italian primary school that follows the Montessori approach. The students, aged 10 and 11, were divided into two classes with 16 and 19 students, respectively. The activity was conducted separately for each class in February 2024. The students were already familiar with concepts like bits and algorithms, thanks to various "plugged" and "unplugged" projects carried out by the school. The goal of the proposed activity was to show that there exist different types of computational problems: some problems can be solved in a short time (easy problems), while some problems can be solved in a very long time (hard problems). In more academic terms, we aimed to introduce the concept of complexity classes. Another objective was to introduce graphs as one of the most important topics in computer science.

The first part of the activity introduced the concept of graph and the SSSP problem as described in Sect. 3. The SSSP problem can be "easily" solved using Dijkstra's algorithm. This part was introduced with references to navigation apps like Google Maps and Apple Maps, which are familiar to students. The second part of the activity focused on "hard" problems and, in particular, on the graph coloring problem, following the approach in [4]. Students realized that even an intuitive problem, such as coloring the vertices of a graph, cannot be "easily" solved, even with small graphs.

The students showed curiosity and actively participated in the challenges presented in both parts; several questions were asked by students, even in the following days. The teachers were positive about the activity and student engagement, and they are interested in repeating the activity next year. We observed that, since the school follows a Montessori approach, the students were accustomed to using physical devices to better understand abstract concepts (e.g., mathematical operations, language, geography) and were curious about the mechanical device used to explain Dijkstra's algorithm.

5 Conclusions

In this paper, we have introduced a mechanical device to explain Dijkstra's algorithm. The device is a board with a graph made of chains and rings connected to the table via a retractable badge holder; by suitably pulling nodes, we have a visual representation of Dijksta's algorithm. As future work, it would be interesting to perform a formal evaluation of the effectiveness of the device in teaching, and to analyze which other graph algorithms can be explained using the proposed device. Moreover, it would be interesting to develop other mechanical devices to visualize other algorithms, for instance for sorting and searching but also for some simple learning methods.

Acknowledgments. Riko Jacob would like to express his gratitude towards all the people at ETH Zurich, TUM in Munich and IT University of Copenhagen, that were involved in designing and creating the different mechanical devices and exhibitions, and for showing them in class and to the public. Francesco Silvestri is grateful to teachers Ylenia Costa and Cinzia Paccagnella of the *Scuola Primaria M. Montessori* (Padua, Italy) for the opportunity to present the activity to their students. Silvestri was in part supported by project "Big Data Analytics for Mobility" under the UniImpresa call of the University of Padua, and by the PRIN project n. 2022TS4Y3N "EXPAND: scalable algorithms for EXPloratory Analyses of heterogeneous and dynamic Networked Data", funded by the Italian Ministry of University and Research (MUR).

References

1. Battal, A., Adanır, G.A., Gülbahar, Y.: Computer science unplugged: a systematic literature review. J. Educ. Technol. Syst. **50**(1), 24–47 (2021)
2. Bell, T., Rosamond, F., Casey, N.: Computer science unplugged and related projects in math and computer science popularization. In: Bodlaender, H.L., Downey, R., Fomin, F.V., Marx, D. (eds.) The Multivariate Algorithmic Revolution and Beyond. LNCS, vol. 7370, pp. 398–456. Springer, Heidelberg (2012). https://doi.org/10.1007/978-3-642-30891-8_18
3. Bell, T., Vahrenhold, J.: CS unplugged—how is it used, and does it work? In: Böckenhauer, H.-J., Komm, D., Unger, W. (eds.) Adventures Between Lower Bounds and Higher Altitudes. LNCS, vol. 11011, pp. 497–521. Springer, Cham (2018). https://doi.org/10.1007/978-3-319-98355-4_29
4. Bell, T., Witten, I., Fellows, M.: Computer science unplugged - an enrichment and extension programme for primary-aged children (2002)
5. Cormen, T.H., Leiserson, C.E., Rivest, R.L., Stein, C.: Introduction to Algorithms, 3rd edn. The MIT Press, Cambridge (2009)
6. Dijkstra, E.W.: A note on two problems in connexion with graphs. Numer. Math. **1**(1), 269–271 (1959)

Unplugged Decision Tree Learning – A Learning Activity for Machine Learning Education in K-12

Lukas Lehner(✉)[iD] and Martina Landman[iD]

TU Wien, Vienna, Austria
{lukas.lehner,martina.landman}@tuwien.ac.at

Abstract. Artificial intelligence (AI) is now deeply ingrained in young peoples' everyday lives. They need low-threshold learning opportunities to understand what AI is and how it works. Unplugged learning activities can offer such opportunities but must first manage to break down the topic's complexities. This contribution presents such an activity giving students hands-on experience in training an actual machine learning (ML) model - all without a computer! 'Actual' here refers to the fact that the model students train ends up being an exact copy of what a standard Python implementation would produce. Three tools are presented that make this feasible in an unplugged two-hour workshop setting. We report our experience piloting the activity including questionnaire responses we collected from 56 upper secondary school students.

Keywords: Machine learning education · Unplugged · Decision tree learning · K-12 education

1 Introduction

With the increasing prevalence and relevance of artificial intelligence (AI) in peoples' everyday lives comes an increased need for AI literacy, especially for young people. Fostering competencies to contribute to this literacy is slowly integrated into school curricula worldwide [11], while what exactly constitutes AI literacy is still under discussion [4]. Until schools can fully provide the needed education, university outreach programs may offer educational activities to fill the gap. Our institution offers workshops to primary and secondary school classes on a variety of computer science (CS) topics. To match the increasing demand for AI literacy, we developed an unplugged learning activity which we present in this experience report. The activity introduces students to AI and machine learning (ML) by way of decision tree learning (DTL). Students first 'handcraft' a fruit-classifying tree based on their own intuition using 3D printed 'tree parts'. Then they train another tree that classifies aliens, but this time they use a formula – a simplified version of the *Gini Impurity* – to calculate which tree part should 'grow' where. The target audience are upper secondary school

© The Author(s), under exclusive license to Springer Nature Switzerland AG 2025
H. Fernau et al. (Eds.): CMSC 2024, LNCS 15229, pp. 50–65, 2025.
https://doi.org/10.1007/978-3-031-73257-7_4

students. Their main takeaway should be a good understanding of what AI is and where the line between ML and non-ML systems lies. Special focus lies on demonstrating how simple rules and formulas can create an intelligent seeming system. In this contribution, we present three key ideas (or tools) that make training DTL models feasible in our setting. They are the quickly adaptable *tree parts*, a *simplified formula* for building the tree, and a novel way of representing the training data using *data cards*, which allow students to quickly and intuitively find the values needed to calculate the formula. These tools allow students to discover and use DTL techniques at a very low level of abstraction (i.e., very close to their original form), which aims to demystify the inner workings of an ML model by revealing how it classifies data and learns to do so from examples.

2 Related Work

Unplugged learning activities are a widely used tool to teach complex computer science subjects to non-experts [1]. Several unplugged activities have been proposed on the subject of AI including various ML paradigms [6, 8–10, 13, 15].

Regarding decision trees, Lindner et al. [6] present an activity, where students create a binary classifier for illustrations of monkey-faces by coming up and writing down decision rules based on the provided labelled training images. Students learn important ML concepts such as training and test data, how a decision tree model classifies data, and that a 100% accuracy is not guaranteed. The activity, however, does not convey a core concept of machine learning: how a machine *learns*. Students are the ones 'growing' the decision tree, choosing decision rules based on their own considerations. These considerations may not be implementable (and thus automatable) on a machine, e.g., if they are solely based on students' intuition. A similar activity (*Pasta Land*) is proposed by Ma et al. [7]. Michaeli et al. [12] go a step further, using the activity by Lindner et al. [6] in their teaching concept's first phase and following it up with the use of a digital data analysis tool. There, students can let their computer build a decision tree based on the monkey data and compare the model's performance to that of their unplugged tree. While the tool allows the specification of stopping and pruning parameters, splitting criteria are not mentioned or explained, which, again, does not provide a clear picture of *how the machine learns*. Our contribution lets students uncover the learning process completely by having them use a formula to calculate best splits, and thus, experience exactly how a machine learns from data.

Podworny et al. [14] developed *data cards* for their introductory unplugged teaching unit on decision trees. Each card shows an image of a different food item and a list of nutritional values, such as the amount of calories per 100 g of the shown item. These data cards allow the discussion and demonstration of a wide range of DTL topics including what a *split* is, but teaching the entire DTL process (including *splitting criteria*) is again delegated to a follow-up lesson using a digital tool (presented in [2]). A drawback of these data cards is that in order to fully perform the DTL process and evaluate every possible split at a given node, the cards would have to be consecutively sorted according to each of the seven numerical attributes - a time-consuming task which does not contribute to AI centred learning goals. Our contribution mitigates this issue by introducing a novel data card design that removes the need to sort cards by (numerical) attributes, making the execution of the full DTL process in an unplugged setting more feasible.

Every contribution we are aware of uses pen and paper to represent the unplugged decision tree model. Drawn trees, however, cannot easily be adapted, e.g., inserting a node near the root of a finished tree or moving whole sections around requires the erasure and redrawing of wide parts of the tree. Also, when starting out and not knowing how big the tree will end up being, students may use up too much space in the beginning, having to then make remaining nodes smaller and smaller to be able to fit them on the page (Fig. 5 in [7] shows an example of this). Our contribution introduces tangible 3D printed tree elements that allow students to quickly adapt their trees.

3 Three Tools to Make Unplugged DTL Feasible

The setting of our learning intervention is an unplugged two-hour workshop for upper secondary students that have no prior knowledge of AI or CS. The workshop's aim is to 'demystify' AI by letting students discover how a formula is used to train an ML model on data. Such an exact training process can be complex and computationally intense and is usually left out of unplugged learning activities or delegated to a computer. Since we want students to discover this 'actual' training process, but our setting lacks computers, we developed the following three tools to make the process feasible anyway.

3.1 A Tangible and Quickly Adaptable Tree

We designed 3D printable 'tree parts', that allow students to construct tangible and quickly adaptable decision trees (see Figs. 1 & 5 for finished examples).

A tree consists of three types of 3D printed parts: *branches* used for inner nodes, *leaves* for leaf nodes and *twigs* for edges of a (binary) decision tree. Stickers are used to label branches with yes-or-no questions (e.g., 'Green?') and leafs with a class (e.g., 'Apple'). Students can relabel branches and leaves to their liking and are thus not bound by the labelled tree parts made available to them. Twigs are printed in two distinct colours, so that one can be used as 'yes'-twig and one as 'no'-twig without having to be labelled with extra stickers. The triangular branches and leaf (or arrow head) shaped leaves have knobs for feet on their bottom side that fit loosely into a twig's rectangular hole. The loose fit of the individual parts allows students to move, twist, and reshape trees and exchange elements easily during assembly. This allows quick prototyping and (re–)modelling of a tree without having to use an eraser, start from scratch, or end up with an unreadable model. We are not aware of any previous contribution using a building block system for decision tree representation like ours. We additionally designed a 'garden' for the students to 'grow' their tree in. It is a A3 sheet of paper displaying a garden with a 3D printed stump shaped object glued onto it. There, students can anchor their first branch, i.e. 'plant' their tree.

Fig. 1. A hand-crafted unplugged decision tree to classify six different fruits. The data cards classified by the model are next to the leaves they ended up at.

3.2 A Simplified Formula

At the heart of decision tree learning lie the rules for when to ask which question (i.e., what inner node to add to the tree) and when to stop (i.e., when to place a leaf node). Thus, when reducing the complexity of DTL to make its main ideas teachable to non-experts in an unplugged environment, these rules need to be simplified. We decided to use the simple stopping criterion *class purity*, where a leaf node is added when only training data from the same class ends up at that node. We decided against any post-pruning techniques and rather designed the fake data set, so that no (unintended) over-fitting would occur. For measuring the quality of a split, i.e., deciding which question to ask, we chose one of the most widely used[1] metrics: the *Gini Impurity*.

A textbook definition (adapted from [5]) of the Gini Impurity metric reads: Let $X = \{x_1, ..., x_n\}$ be the set of all training data and $S = \{1, ..., k\}$ a splitting rule, then S divides X into a set of subsets $\{X_1, ..., X_k\} =: X_S$, where k is the number of child nodes of an internal node. Let $Y = \{y_1, ..., y_m\}$ be the set of class names, i.e. labels, then $|X_{ij}|$ is the number of occurrences of class j in

---

[1] It is used as default by scikit-learn, a popular Python library for ML.

subset X_i. Let $p_{ij} = \frac{|X_{ij}|}{|X_i|}$ be the ratio of members of class j to the whole subset. Then the *Gini Criterion* of a subset can be defined as

$$Gini(X_i) := \sum_{y \in Y} p_{iy}(1 - p_{iy})$$

The *Gini Impurity* of a splitting rule S is then defined as

$$F_{Gini}(S) := \sum_{i \in S} \frac{|X_i|}{|X|} Gini(X_i)$$

The sight of such definitions and formulas may, of course, discourage some students. Thus, we present a simplified, more user-friendly version of the Gini Impurity: First we decide to use a binary decision tree and only two classes and can thus substitute $k = 2$ and $m = 2$. This already removes complexity and gives us $X = X_1 \cup X_2$, $X_1 = X_{11} \cup X_{12}$ and $X_2 = X_{21} \cup X_{22}$. We label the cardinalities of the subsets as follows: $a := |X_{11}|$, $b := |X_{12}|$, $c := |X_{21}|$, and $d := |X_{22}|$. These four (integer) variables are the only ones the students will have to deal with. We continue substituting and simplifying:

$$Gini(X_1) = \frac{a}{a+b}\left(1 - \frac{a}{a+b}\right) + \frac{b}{a+b}\left(1 - \frac{b}{a+b}\right)$$
$$= \frac{ab}{(a+b)^2} + \frac{ba}{(a+b)^2} = \frac{2ab}{(a+b)^2}$$

and similarly $Gini(X_2) = \frac{2cd}{(c+d)^2}$. From this immediately follows

$$F_{Gini}(S) = \frac{|X_1|}{|X|} \cdot Gini(X_1) + \frac{|X_2|}{|X|} \cdot Gini(X_2)$$
$$= \frac{a+b}{n} \cdot \frac{2ab}{(a+b)^2} + \frac{c+d}{n} \cdot \frac{2cd}{(c+d)^2}$$
$$= \frac{2}{n} \cdot \left(\frac{ab}{a+b} + \frac{cd}{c+d}\right).$$

The leading fraction $\frac{2}{n}$ is the same for all splits at a node. Thus, it will not impact their ranking and can be omitted, giving us the *Simplified Gini Impurity*:

$$F_{SimpleGini}(S) := \frac{ab}{a+b} + \frac{cd}{c+d}.$$

Since a, b, c, and d are integers, this formula relies solely on maths operations a high school student should be able to perform with pen and paper. However, given that the focus of the proposed learning activity is ML and not maths, and that the formula will have to be calculated many times over, the use of a pocket calculator is advised. This breaks the 'no computers' rule of classic CS Unplugged activities, however, given that pocket calculators are rather low tech, easy to come by in a school environment and only perform a task students *could* do themselves (and thus fully understand), we believe its introduction violates the rule only in a very minor way. Other metrics such as the *DKM Criterion, Error Rate*, or *Information Gain* can be reduced in a similar way, have however shown to either be still too complex after reduction (e.g., by containing the logarithmic function) or not perform well enough on the fake data sets we created. To further facilitate the use of the simplified formula, we designed a help sheet that explains visually how to extract the four variables a, b, c, d from a given split (Fig. 2).

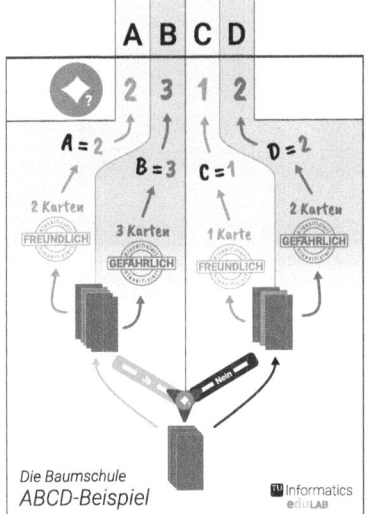 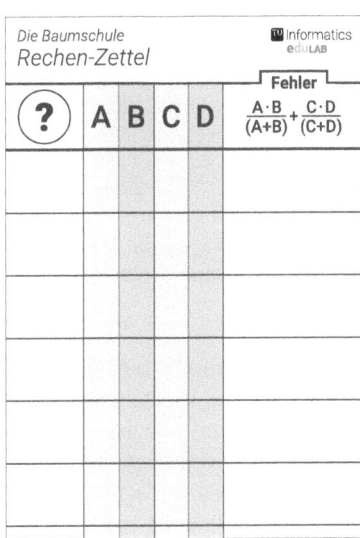

Fig. 2. Left: A help sheet demonstrating by example how to extract the four variables needed to calculate the *Simplified Gini Impurity* of a question. Right: A sheet for executing the calculation for multiple questions.

3.3 A Novel Data Card Design

In order to fully train an unplugged decision tree the simplified formula presented above may have to be used very frequently: Let $X = \{x_1, ..., x_n\}$ be the subset of data for which we search the best split. Given an attribute with $k \leq n$ different values occurring in X, $k-1$ thresholds can be drawn between those values (although this can be optimised by skipping class clusters). Let a be the number of attributes, then in the worst case the formula needs to be calculated $an - 1$ times, which is just to add *one* branch to the tree. One way to reduce the number of calculations needed is to keep a and/or n small. This, however, would force the use of small, low complexity data sets, which in our opinion works to diminish the effect we are trying to evoke in students: awe at how a simple formula can outperform humans and grow a better decision tree faster

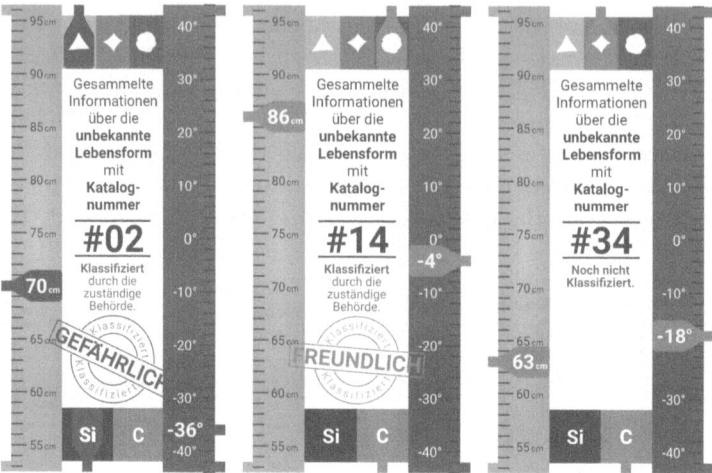

Fig. 3. Three data cards from the Aliens data set, labelled as 'dangerous' (left), 'friendly' (centre) and a test data card not yet classified (right).

than the students themselves could. If the data set is too simplistic, students may find patterns in the data and grow a tree faster without the formula.

We investigated how to speed up the process of finding the best split by other means. We observed that calculating the formula is a two-step process: find the appropriate values for a, b, c, and d, then execute the calculation. The latter is streamlined by allowing the use of pocket calculators. The former usually involves the sorting of the data in X according to one attribute, then choosing a value and counting how many data points of each class are on each side of that threshold. Then the threshold is moved to the next occurring value (or into the space between two values) and the class frequencies are counted again. This is repeated for all possible thresholds before moving on to the next attribute and sorting the data anew. In order to sort data in an unplugged setting, the individual data points can, for example, be represented as physical, movable objects. Playing card style *data cards*, such as those presented in [14], immediately come to mind. They seem well suited to this task. However, since every attribute requires a full sorting of the data (i.e., deck of cards), valuable workshop time will be spent on the sorting processes, which themselves are not of interest here. Thus, we developed a novel data card design that extensively reduces the need for sorting (Fig. 3): each side of the rectangular cards corresponds to one attribute. There are two numerical attributes measured in 'cm' (left side) and '◦' (right side), a categorical attribute that is either 'Si' or 'C' (bottom), and a categorical attribute that can be one of three shapes (top). We want students to solely rely on the data presented when training a tree (just like a machine would) and not infuse their own potential background knowledge into the system. This is why the attributes do not have names or any intended meaning. The data should feel alien to the students (hence the creation of an *Aliens* data set including a story line of saving the world). The value for each attribute is indicated by a tag on the side of the card. Each tag has a part that sticks out over the edge of the card like a knob. These knobs are visible from above even when multiple data cards are stacked on top of each other (Fig. 4). Only knobs representing the same value overlap, which is why we manufactured the data set so that numerical values occur at most once per attribute while retaining the intended correlations.

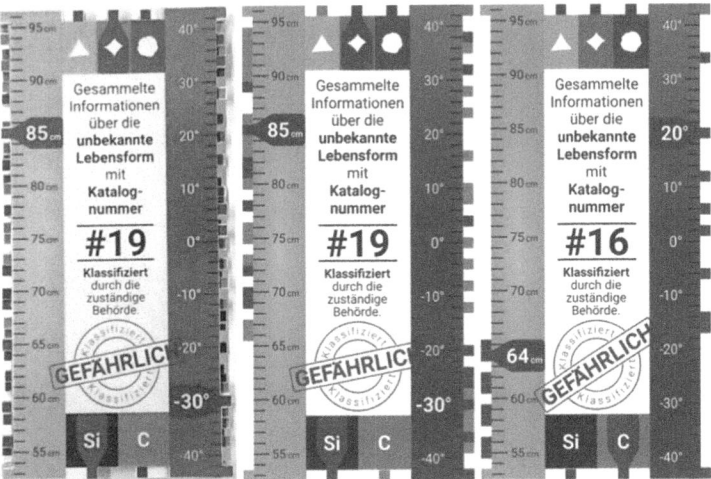

Fig. 4. A photo of a stack of all Alien training data cards (left) and renderings of the same stack but split in two using the first best split '66 cm or more' (centre & right).

The class label of a data card is indicated by text (stamp like text on the bottom half of the card) and colour (red or green colour of the stamp, its text and the four tags). Test data cards do not have a stamp and their tags are blue. Their true class label is written in 'invisible ink' where the stamp is for training data cards. This label is only visible under UV light (Fig. 5 bottom left). With this card design, the process of determining values for $a, b, c,$ and d has been reduced to stacking the data cards and counting off the occurrences from tags sticking out on the left or right side for the left or right attribute respectively. For categorical attributes the deck is best sorted by hand (in $\mathcal{O}(n)$) into one pile per category, at which point the class frequencies can be read off of the left or right side knobs.

4 'The Tree Nursery': A Learning Activity

We developed a learning activity that utilises the three presented tools to let students discover how a decision tree is trained from data. In a first phase students are familiarised with the materials: they discover how to assemble a decision tree from the tree parts and use data cards to test their model. In the second phase they figure out how to use the data cards in combination with the *Simplified Gini Impurity* to let 'the tree grow on its own', instead of assembling it using their intuition. Figure 5 shows a typical end result of the second phase. A detailed description and a rundown of the activity's two main phases can be found in Appendix A.

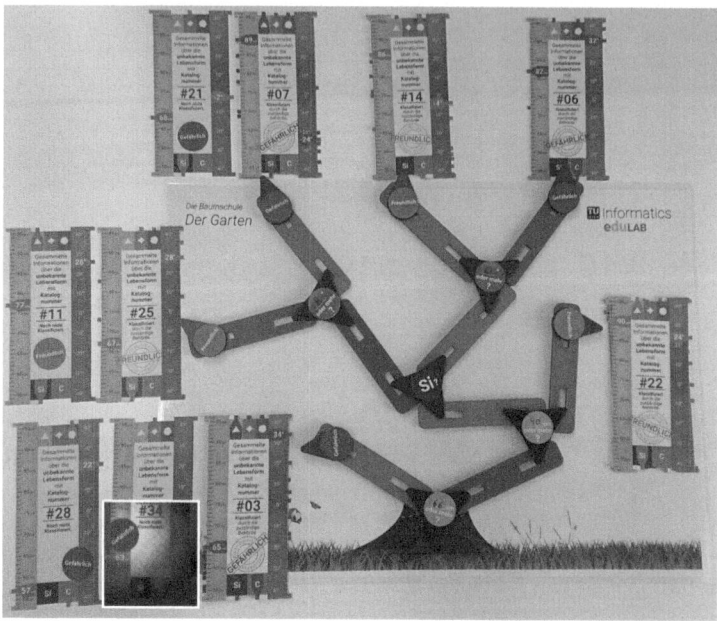

Fig. 5. An unplugged DTL model trained on the Aliens data set using the Simplified Gini Impurity. Also shows training and test data cards at the leaves they end up at when classified by the model. On the bottom left is the designed-to-be-misclassified test data point 'Alien #34' (classified as 'dangerous' while a UV flashlight reveals it is actually 'friendly').

4.1 Piloting the Learning Activity

During development and testing of the activity, 32 workshops with in total more than 650 students were conducted successfully. We observed that students are easily able to construct their first decision tree using the tree parts and written instructions. Some students add too many leaves to twigs. When asked to assemble a tree that classifies all data correctly, some students need hints as to how to strategically construct the tree. Most, however, discover this on their own. Students understand quickly how to use the data cards to count off the values needed for the formula. However, the cards themselves turned out to be very labour-intensive to produce. The formula seems simple enough for students to calculate using a calculator or, in easy cases, by hand. The data sets' themes (fruits & aliens) seem to be engaging and the alien story line seems motivating enough for students to eagerly calculate the formula multiple times.

For four workshops we collected data using two questionnaires (one right before and one right after each workshop). Of 68 responses, 56 were filled out completely (9 did not fill out the second questionnaire) and seriously (3 had same answers on all Likert items), and we report on these here. Both questionnaires had participants respond to Likert scale items (see Appendix B) relating to AI literacy, such as "I know definitions of 'artificial intelligence'.". Items were selected and slightly adapted from [3]. Possible ratings of each item were 'Fully applies', 'Rather applies', 'Rather does not apply', 'Does not apply', and 'Don't know'. All items are formulated so that 'Fully applies' implies the highest level of AI literacy. The goal here was to detect change in participants' self-assessment of their AI competencies from before to after our learning intervention. The questionnaire before the workshop also asked for participants' gender, the one after posed an optional open-ended question: "Here is space for your comments, feedback, and criticism! How did you like the workshop?". The goal here was to gather some general feedback and get a sense of participants' attitude towards the workshop. Table 1 shows the number of participants whose responses were included in our analysis, their gender and the grade they were in.

Table 1. Demographic data of participants included in the analysis.

Workshop	Grade	Male	Female	Other	**Total Responses**
1	12	6	4	1	11
2	10	11	3	1	15
3	10	9	6	1	16
4	9	5	7	2	14
Total Responses		31	20	5	56

Figure 6 visualises changes in Likert scale ratings from before to after a workshop. We observe a general upward trend from lower confidence in students' own AI competencies (we regard 'Don't know' as lowest rating) to higher confidence: 48.75% of ratings improved. 45% stayed the same, a third of which (15% of the total) stayed at the highest rating. 6.25% decreased. The latter we posit might include students realising how little they knew about the subject, and had previously overestimated their competence.

We coded answers to the open-ended question using a 5-star rating system: 5 stars for answers describing the workshop as 'very good', 4 stars for 'good', and 3 stars for 'OK' (which were the lowest rated answers). Using this rating system the workshop received an average of 4.55 stars. 21,42% of participants did not answer this optional question, 7,14% did not provide an answer interpretable as rating. One student noted the workshop was "A little to long". Highlights of

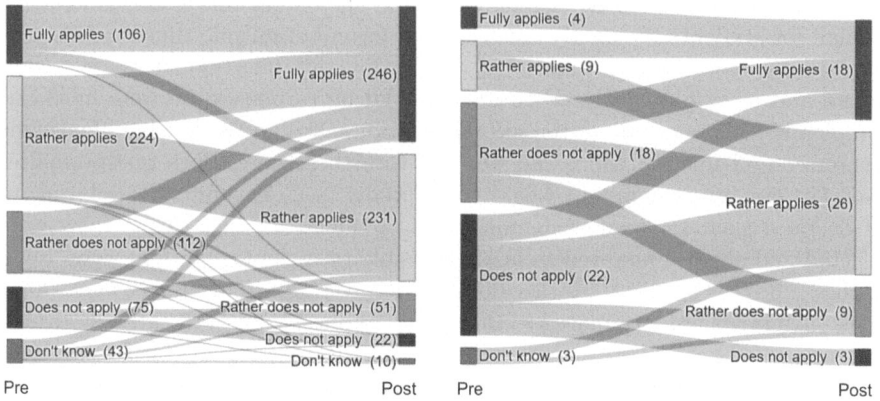

Fig. 6. Sankey diagrams showing changes in student self-assessment from before to after a workshop. Left: combined ratings of all Likert items. Right: ratings of the item "I can design new AI applications". Both diagrams show data from all four workshops combined. Note that the highest rating ('Fully applies') overall more than doubled.

students' answers include (translated from German): *"It was a cool workshop and I really enjoyed it. Good explanation"*; *"Very well thought-out unit, exercises illustratively designed, practical learning of AI: very positive"*; *"It was very nice, very exciting, and interesting new topics that you didn't know about yet"*.

5 Conclusion

This contribution presented three novel tools to make unplugged learning activities on decision tree learning feasible. During piloting of our own learning activity utilising these tools we could successfully observe this feasibility. Students easily constructed decision trees from the 3D printed *tree parts* and were able to quickly adapt them as needed. The *data cards*, while being labour-intensive to produce, enabled students to streamline the process of training a model by letting them quickly count off the values needed for the *Simplified Gini Impurity* formula. This formula appears to be simple enough for students to calculate quickly and not deter them from being engaged in the activity, while still producing the same models as a standard Python implementation would.

We plan to develop follow-up materials for the classroom in an attempt to improve long-term effects as reported in [1]. All materials and files needed to run our activity will be made available under a Creative Commons license here: https://edulab.ifs.tuwien.ac.at/materialien/ai_dtl/.

Acknowledgments. This work was partially funded by LEA Project 2WP-0073 on 'Abenteuer Informatik' & WWTF support for female/diversity promotion at TU Wien.

Disclosure of Interests. The authors declare that they have no conflict of interest.

A Details on 'The Tree Nursery' Learning Activity

A.1 Learning Objectives

When having completed the activity 'The Tree Nursery', students will be able to ...

1. Describe and demonstrate how a decision tree classifies a data point.
2. Construct a classification decision tree based on a small data set.
3. Use a decision tree to classify data.
4. Discuss and explain why a data point is misclassified by a decision tree.
5. Discuss quality aspects of training data.
6. Adapt a decision tree to fit new data.
7. Understand that some features may be more important that others when classifying data and discuss which features may be important when classifying everyday objects.
8. Execute an algorithm (top-down induction of a decision tree) to train a machine learning model (classification tree) based on a small data set.
9. Explain how a decision tree learns from data.
10. Explain why outliers may be misclassified by a decision tree.
11. Name examples of domains in which decision trees are used and for what purpose.

A.2 Activity Rundown

The activity can be done in a classroom setting with a recommended ratio of 1 teacher to 12 students, which are sat in groups of 4. The activity switches between group-work and plenary discussions. During group-work the teacher acts as a facilitator rather than teacher, standing by for questions, and sometimes prompts to help students reflect on their actions so far. The plenary discussions are lead by the teacher. The activity has two main phases: In Phase 1 a decision tree is 'hand crafted' and in Phase 2 an ML method is used to calculate a decision tree.

Phase 1: The Decision Tree

Plenary Discussion: What is AI? After groups of on average 4 students are formed a plenary discussion begins with the question "Anybody want to describe what they think artificial intelligence is?". After some discussion the same question but about 'machine learning' is posed. The goal of the discussion is not for the teacher to provide an answer but to explore existing conceptions among the students. The answer will be discovered throughout the activity. If a correct answer is given by a student, that is alright as well.

Group-Work: Your First Decision Tree. Each group receives the same task sheet with four subtasks to be solved by each group separately:

1. Students receive a bag with tree parts, a garden and an algorithm for assembling a decision tree that they are tasked to execute.
2. Students receive training data cards of a *Fruits* data set (various fruits with 4 attributes each) and an algorithm for using a decision tree to classify a data card. They are tasked to execute the algorithm for all data cards. Students discover that since the tree was assembled randomly, its accuracy is low.
3. Students are tasked to remodel their tree, so that is achieves 100% accuracy on the training set (Fig. 1 shows an exemplary result).
4. Students receive test data cards and the task to classify them using their tree. Some test data is designed to be misclassified for various different reasons.

Plenary Discussion: Why were Some Cards Misclassified? Once all groups finished the task sheet a plenary discussion is held on why some test data cards failed. This leads to a discussion on quality aspects of data and how to handle them. For example, since colour may have been used as deciding factor for the *yellow* banana in the training set, the *red* banana used during testing will likely be misclassified. This spawns a discussion on bias and how the students manifested their wrong conception of a banana always being yellow into their AI system, which now also displays this bias. Or the *squirrel* (attributes: red, seedless, grows on trees, no edible peel) will be used to discuss that decision trees don't really 'recognise' fruit: they only classify input data into available categories, no matter whether the input was actually one of those categories.

Phase 2: The Self-learn Tree

Plenary Discussion: What is ML? The teacher asks whether the trees grown during the first phase should be considered AI, and if so, also ML. The discussion should lead to the consensus that the trees were AI but not ML, because there was no learning done by a machine, but rather the students used their knowledge and reasoning skills to create their AI system. Now the use of a formula, the *Simplified Gini Impurity*, is motivated: We want to automate the tree growing process. It is too much work for us humans, we want a computer to do it for us. But what is the only thing a computer can do? Calculate. Thus we have to find a way to calculate a decision tree.

Group-Work: Finding the Best Split. This phase is done cooperatively by the whole class. The teacher introduces the story: *ALIENS!!* are heading towards earth and students need to grow a decision tree to determine which aliens can be trusted and which should be repelled. A set of 30 data cards describing aliens using four attributes and a classification into 'friendly' and 'dangerous' are available (from previous invasions). These training data cards are distributed among groups including copies of the ABCD-example and calculation sheet shown in Fig. 2 with the task of figuring out how to calculate the formula. This might be a challenge for some groups and can be aided by guiding them towards the answer or handing out the solution in form of an algorithm sheet, but experience has shown it to be more rewarding if students discover the mechanics themselves. Once every group knows how to calculate the formula, they are tasked with calculating the 'Error', i.e. impurity, of 3 to 4 questions, for example '66 cm or more?'. It is made clear to the students that a computer would calculate the impurity for every possible question, not just 3 or 4. Every group announces the lowest impurity they found. If the question with the actual lowest impurity was not among those analysed by the students, the teacher has to act as if they also calculated some questions and found an even better one. Either way, a branch with the best question on it is placed into one shared tree in the centre of the room. The training data are split according to the best question. One side of the split will contain only dangerous aliens, thus a leaf is placed in the tree. For the other side the process continues with groups calculating the next best question. This repeats until the tree has no more open twigs. Now the test data cards, i.e., unknown aliens on their way to earth, are handed out and classified by students. After some dramatic drum rolls the true intentions of each alien is revealed using a UV flashlight to reveal its class written in invisible ink on each test card. One alien will be misclassified, spawning a discussion on the concept of outliers and trust in ML models. Figure 5 shows a typical end result.

Closing Plenary Discussion. A final discussion is kicked off with the old question that students now should answer with a resounding 'yes': "Is the tree we just grew AI? Is it an ML model?". Then the exact difference to the first tree and the role humans, i.e. the students, played in both these models is investigated. In the following several aspects of DTL and ML in general can be discussed, such as explainability vs. black boxes or domains of application. But also in-depth discussions are possible, e.g., on why the formula we used works and whether there are alternatives, what other optimisation strategies exist, and which of these might have helped to not misclassify that one friendly alien.

B Questionnaire - Likert items

1. I know the most important concepts of the topic "artificial intelligence".
2. I know definitions of "artificial intelligence".
3. I can assess the possibilities of using an AI application and where its limits lie.
4. I can assess the advantages and disadvantages of using an AI application.
5. I can come up with new application possibilities for AI-based applications.
6. I can weigh the consequences of using AI applications for society.
7. I can make ethical considerations and include them when deciding to use data provided by an AI application.
8. I can analyze AI applications for their ethical implications.
9. I can design new AI applications.
10. I can program new AI applications.

References

1. Bell, T., Vahrenhold, J.: CS unplugged—how is it used, and does it work? In: Böckenhauer, H.-J., Komm, D., Unger, W. (eds.) Adventures Between Lower Bounds and Higher Altitudes. LNCS, vol. 11011, pp. 497–521. Springer, Cham (2018). https://doi.org/10.1007/978-3-319-98355-4_29
2. Biehler, R., Fleischer, Y.: Introducing students to machine learning with decision trees using CODAP and Jupyter notebooks. Teach. Stat. **43**(S1), S133–S142 (2021). https://doi.org/10.1111/test.12279
3. Carolus, A., Koch, M.J., Straka, S., Latoschik, M.E., Wienrich, C.: MAILS - meta AI literacy scale: development and testing of an AI literacy questionnaire based on well-founded competency models and psychological change- and meta-competencies. Comput. Hum. Behav.: Artif. Hum. **1**(2), 100014 (2023). https://doi.org/10.1016/j.chbah.2023.100014
4. Casal-Otero, L., Catala, A., Fernández-Morante, C., Taboada, M., Cebreiro, B., Barro, S.: AI literacy in K-12: a systematic literature review. Int. J.STEM Educ. **10**(1), 29 (2023). https://doi.org/10.1186/s40594-023-00418-7
5. Lee, V.E., Liu, L., Jin, R.: Decision trees: theory and algorithms. In: Data Classification: Algorithms and Applications. Chapman and Hall/CRC (2014)
6. Lindner, A., Seegerer, S., Romeike, R.: Unplugged activities in the context of AI. In: Pozdniakov, S.N., Dagienė, V. (eds.) ISSEP 2019. LNCS, vol. 11913, pp. 123–135. Springer, Cham (2019). https://doi.org/10.1007/978-3-030-33759-9_10
7. Ma, R., Sanusi, I.T., Mahipal, V., Gonzales, J.E., Martin, F.G.: Developing machine learning algorithm literacy with novel plugged and unplugged approaches. In: Proceedings of the 54th ACM Technical Symposium on Computer Science Education, SIGCSE 2023, vol. 1, pp. 298–304. Association for Computing Machinery, New York (2023). https://doi.org/10.1145/3545945.3569772
8. McOwan, P., Curzon, P.: The brain-in-a-bag activity (2014). https://teachinglondoncomputing.org/resources/inspiring-unplugged-classroom-activities/the-brain-in-a-bag-activity/
9. McOwan, P., Curzon, P.: The intelligent piece of paper activity (2014). https://teachinglondoncomputing.org/resources/inspiring-unplugged-classroom-activities/the-intelligent-piece-of-paper-activity/
10. McOwan, P., Curzon, P.: The sweet learning computer (2016). https://teachinglondoncomputing.org/the-sweet-learning-computer/
11. Miao, F., Shiohira, K.: K-12 AI curricula. A mapping of government-endorsed AI curricula. UNESCO Publishing **3**, 60 (2022). https://unesdoc.unesco.org/ark:/48223/pf0000380602
12. Michaeli, T., Seegerer, S., Kerber, L., Romeike, R.: Data, trees, and forests - decision tree learning in K-12 education. In: Proceedings of the Third Teaching Machine Learning and Artificial Intelligence Workshop, pp. 37–41. PMLR (2023). https://proceedings.mlr.press/v207/michaeli23a.html
13. Ossovski, E., Brinkmeier, M.: Machine learning unplugged - development and evaluation of a workshop about machine learning. In: Pozdniakov, S.N., Dagienė, V. (eds.) ISSEP 2019. LNCS, vol. 11913, pp. 136–146. Springer, Cham (2019). https://doi.org/10.1007/978-3-030-33759-9_11

14. Podworny, S., et al.: Using data cards for teaching data based decision trees in middle school. In: Proceedings of the 21st Koli Calling International Conference on Computing Education Research, Koli Calling 2021, pp. 1–3. Association for Computing Machinery, New York (2021). https://doi.org/10.1145/3488042.3489966
15. Virtue, P.: GANs unplugged. Proceedings of the AAAI Conference on Artificial Intelligence, vol. 35, no. 17, pp. 15664–15668 (2021). https://doi.org/10.1609/aaai.v35i17.17845

Tactile Kolam Patterns – Communicating Art and Mathematics to Students with Vision Impairments

Robinson Thamburaj[✉], Krishnamachari Desikachari, and Gnanaraj Thomas

Madras Christian College, Chennai, India
robinson@mcc.edu.in

Abstract. Kolam is a south Indian artform of drawing patterns on the ground in front of houses for decoration purposes. Relating these cultural designs as illustrations in mathematics classroom facilitates the learning enjoyable. This paper connects the aspects such as (i) representation and exploration of kolam patterns in terms of kolam moves that are similar to Turtle moves, (ii) accessibility of kolam patterns for vision impaired people as tactile diagrams (iii) relevance of a few elementary mathematical concepts and (iv) expression of simple kolam patterns with an inclusive tool, Tactile Kolam Cube. With an ethnomathematics perspective this paper provides scope to appreciate the cultural aspect on one hand and the mathematical properties on the other. We propose the tactile kolam sheets as teaching aids and tactile kolam cube as an assistive tool, and hence the focus of this paper is on these as classroom resources in mathematics pedagogy. Two types of kolam patterns viz. Kambi kolam and Hridhaya kamalam are referred in this paper.

Keywords: Ethnomathematics · Tactile geometry · Assistive device · Kolam patterns · Mathematics education

1 Kolam the Traditional Artform

Like the artform Sona sand drawing traditionally practiced in central Africa, Angola and Congo, the kolam is a traditional Indian folk art widely used to decorate the ground in front of the houses every morning as a gesture to invite prosperity or welcome deities. This decoration of the floor with aesthetic kolam patterns is carried out by women who deftly draw with pinches of rice flour or millet flour. It is thought to be an act of goodwill to begin a day by feeding ants and small birds with rice flour. To draw kolam, initially a lattice of dots is drawn. Holding the flour between thumb and first finger and letting the powder fall in a continuous line, by moving the hand a kolam pattern is drawn. A kolam could be drawn as a single, unsegmented, closed thread of line or superimposition of two or more closed threads of lines, each constituting one component of the global kolam pattern. During festive occasions and days of religious significance, the kolam designs are more elaborate and complicated. Detailed study on the cultural aspects of Sona drawings, kolam patterns has been done [2–4, 6, 9, 14, 19]. Pattern drawn with colorful primitive designs known as Rangoli is color variant of kolam patterns.

2 Representation of Kolam Patterns

Kolam pattern can get complex with many interwoven lines or primitive designs, but they all are drawn on a grid or an array of dots – the grid being a square, hexagon, circle and other free shapes. Smooth and continuous lines are drawn joining the dots or drawn in between the dots, but not retracing the lines once drawn. Kolams can be largely classified into three types - patterns where the lines go between or around the dots, (*kambi* kolam, meaning a string), patterns where the lines connect the dots, and patterns that connect dots and include picture primitives such as lamps, flowers, animals, etc. (Figure 1).

Fig. 1. (a) A kambi kolam (b) A kolam pattern joining dots and primitives

2.1 Kambi Kolam Patterns

Initially an array or a lattice of dots are drawn on the ground. Starting from a position between two dots straight lines and curves are drawn making angular and smooth turns surrounding the dots as shown in figure 2. Viewing kolam patterns as two-dimensional cycle pictures a syntactic method was advocated [13] and kolam patterns were generated. Like turtle moves on a square grid, cycle rewriting grammar rules were explored [10, 11]. Syntactic methods for generation of arrays or patterns using more advanced schemes were discussed in [11, 18]. As a kambi is a closed curve with or without loops represented in the form of a cycle it can be represented as a sequence of turtle moves along the grid lines. The kambi kolam pattern on a square grid can be represented with a set of kolam moves $K = \{F, R(1), R(2), R(3), L(1), L(2), L(3)\}$ where F indicates for "move forward one unit", R(1) represents a "half a turn to the right", R(2) represents a "a U-turn to the right", R(3) a "complete loop to the right", and similarly for L(1), L(2), and L(3). The picture description for each symbol and a closed kambi are shown in Figure 3 (Source: Siromoney, G., Siromoney, R., Robinson, T, 1989).

The first few members of an aesthetically pleasing family of kambi kolam patterns is expressed as $\{(F^{2n}R(2)F^{2n}L(2))^n F^{2n}R(3)(F^{2n}L(2)F^{2n}R(2))^n L(3)/n > 0\}$ in Fig. 4 (Source: Siromoney, G., Siromoney, R., Robinson, T, 1989). Patterns over the subset of kolam moves, $K' = \{F, R(1), R(2), L(1), L(2)\}$ can generate rectangular mono-lineal Sona patterns. If a rectangle does not give a mono-lineal Sona, then it takes gcd(m, n) lines to draw the same.

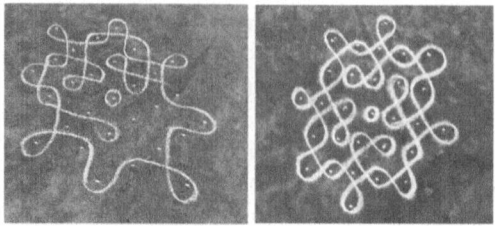

Fig. 2. A) An intermediate stage and (b) a completed kambi kolam

Fig. 3. (a) Kolam moves and (b) a sequence of moves to make a closed cycle

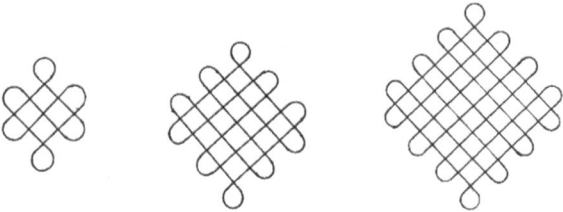

Fig. 4. Three members of the family of kambi kolam patterns.

2.2 Hridhaya Kamalam Kolam Patterns

Unlike the kambi kolam, this type of patterns is drawn by connecting the dots with lines. A special class of such patterns known as Hridhaya kamalam (meaning Lotus Heart) are drawn connecting dots placed on a circular grid. Initially dots are drawn along a set of converging radial arms or axes. Each arm has a fixed number of dots. Once drawn the dots resemble circles and radial arms. Dots arranged in five circles and eight arms shown in Fig. 5 a. A tracing sequence of numbers is used to draw edges connecting dots jumping along the arms clockwise. Figure 5 b shows how the tracing sequence <1,3,5,2,4> starts at a dot in the innermost circle (1st circle) and jumps to subsequent arms connecting the dots at 3rd, 5th, 2nd and 4th circles for every jump clockwise. The sequence is repeated until the line reaches the point where it started. The completed Hridhaya kamalam pattern denoted as HK(8,5,<1,3,5,2,4>) is shown in Fig. 5 c. By varying the number of arms and dots and choosing different tracing sequence patterns can be obtained. In general a Hridhaya kamalam pattern on m arms and n dots with tracing sequence $<s_1, s_2, ..., s_n>$ is denoted as HK($m, n, <s_1, s_2, ..., s_n>$). For the same values of m and n but taking the reversal of the tracing sequence as $<s_n, s_{n-1}, ..., s_1>$ the mirror reflection HK($m, n, <s_1, s_2, ..., s_n>$) is obtained. It can also be noted that the number of closed lines that make up the kolam depends on the greatest common divisor (gcd) of *m* and *n* [15] (Fig. 6).

Fig. 5. (a) Eight arms of dots (b) joining dots clockwise and (c) a completed pattern

Fig. 6. (a) Axial symmetry (b) HK(8, 5, <1,3,5,2,4>) and its mirror image with reversal of tracing sequence

3 Elementary Concepts Through Kolam

3.1 Programming Commands

In the school curriculum of teaching programming at beginner level, the turtle move commands are important primitive steps to understand the drawing of graphics. The attributes such as current position, angle of turn, location, and pen up/down, and the sequence of commands, etc. are essential and logical for understanding the programming language.

At high school level, the students get to explore the pen commands and angle choice to move along a triangular grid and create triangular, hexagonal patterns. In a similar manner the intricate and complex Kolam patterns are expressed in terms of kolam moves. The kolam commands are ideal to introduce concepts such as loops, functions, conditions and create closed cyclic patterns.

3.2 Mathematics Concepts

While introducing geometry at beginner level, the kolam drawn as triangular, rectangular, hexagonal patterns are suitable for introducing polygonal shapes. Further various forms of symmetry – rotational, axial, etc, are well expressed with kolam patterns.

To illustrate or experiment concepts such as gcd of two positive integers, divisibility of a number, relative prime numbers, patterns drawn with K' moves are found to be relevant. The kambi and Hridhaya kamalam kolams are appropriate to demonstrate gcd [4, 15], cyclic permutation, and group theory. For example, HK(5,4,<1, 2, 4, 3>) has a single strand connecting all dots while HK(5,5,<1, 3, 4, 2, 5>) has five strands.

4 Tactile Kolam Patterns in Pedagogy

Kolam patterns are geometrical figures coupled with certain mathematical concepts as mentioned above. Making mathematics accessible to the vision impaired (VI) students is a challenge due to the non-linear information and discrete nature of data. Tactual shape perception enriches the understanding of the geometrical concepts and the graphicacy skill varies with respect to every student's thinking [1]. It is essential to provide appropriate assistive devices, tools, and technologies to meet the student's unique learning needs [13]. The classroom assistive devices with hands-on experience provide greater participation, give a sense of satisfaction resulting in an improved academic performance of disabled students. Hence, there is a need in making assistive devices for accessibility and ensuring the inclusivity for a main stream education.

4.1 Tactile Kolam Sheets

Geometrical or any picture contents pose a challenge to VI students and tactile graphics deals with preparation of raised line diagrams and Braille notations for better tactile perception. [1, 17]. The students with vison impairment encounter a challenge as kolam diagrams drawn on the ground by flour or powder get erased while accessing with fingers. Prepared in swell sheets as tactile diagrams these patterns become accessible for them. Microcapsule papers (also known as swell papers) are widely used to create raised line pictures, tables, and graphs. There are heater machines that process the swell paper and make the lines drawn pop up and tactile enough. Kolam patterns when made as raised line diagrams on swell sheets make kolam accessible for VI persons (Figure 7). Stages of geometry construction presented as progressive steps help the vision impaired student to understand the drawing procedure through hand exploration [12]. Similar approach of progressive step method in teaching kolam drawing procedure of kambi kolam and Hridhaya kamalam patterns was found to be easier for better understanding of the intermediate or developmental stages.

4.2 Tactile Kolam Cube

The tactile kolam sheets not only give access but also help them explore and understand the intricate design and intermediate stages of drawing involved in it. This accessibility leads to appreciate the mathematical elements of the kolam patterns. Once the VI students get familiar with reading tactile kolam sheets they would want to express new patterns. Any kambi kolam is expressed with combining six basic primitive patterns, viz., circle (white), diamond (yellow), tear drop (green), saddle (blue), and eye (pink) as shown in Fig. 8 a. Wooden cubes with embedded tactile primitives as tactile lines were designed

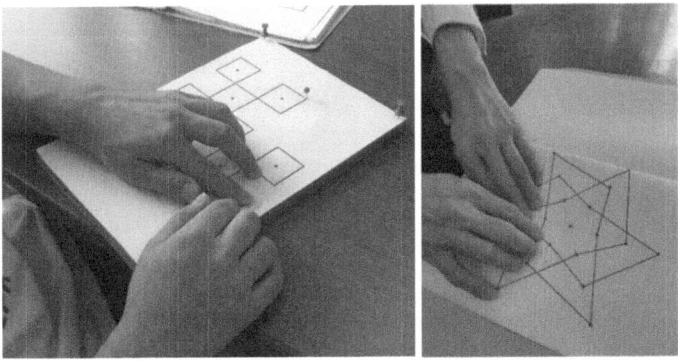

Fig. 7. Hand exploration of tactile kolams on a swell paper by VI persons

Fig. 8. (a) Six primitive patterns (b) Tactile kolam cubes (c) Kolam formation

[7] facilitating the VI people to express kolam pattern formations desired (Figure 8 b, 8 c).

At the dot position of each primitive a strong magnet is inserted. The primitives diamond and circle are embedded on opposite sides, similarly, eye and saddle and tear drop and fan are embedded opposite to each other. When two cubes are brought closer, they get attached due to magnetic attraction. VI persons found it convenient as the cubes do not slip out of hands. By suitably attaching primitives kolam patterns of various sizes are possible. The challenge would be to form a three-dimensional cuboid with closed connected lines on the surface. This cube is an inclusive tool - colored primitives make it attractive for all children [8] and tactile primitives suit children with vision impairment. In the discipline of special needs education, this kit known as PsyKolo has been used for pattern formation by children with learning disability [7] (Fig. 9a, Source: Nagata, Robinson 2006).

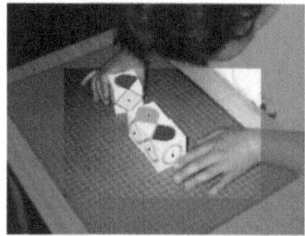

Fig. 9. (a) A child with learning disability exploring (b) Children arranging kolam blocks

5 Conclusion

These tangible tactile materials make kolam patterns accessible to VI students, ideal to communicate mathematical, cultural elements and suitable for inclusive education environment (Fig. 9b, Source: Nagata, 2015). This article has discussed how the cultural aspects of kolam artform can be appreciated in an inclusive educational environment, and the mathematical concepts of the school curriculum can be made relevant. The ethnic and cultural elements of the kolam patterns help in the development of intellectual, social, and emotional aspects of a student apart from imparting knowledge. The kolam pattern can be extended to relate with knots and braids. This study can give scope for relating kolam artform with other disciplines and science concepts in school curricula.

References

1. Argyropoulos, V.S.: Tactual shape perception in relation to the understanding of geometrical concepts by blind students. British J. Vis. Impairment **20**(1), 7–16 (2002)
2. Ascher, M.: Ethnomathematics - A Multicultural View of Mathematical Ideas. Taylor & Francis (1991)
3. Ascher, M.: The kolam tradition: a tradition of figure-drawing in Southern India. Am. Sci. **90**(1), 56–63 (2002)
4. Chavey, D.: Mathematical Experiments with African Sona Designs, Bridges: Mathematics, Music, Art, Architecture, Culture, pp. 305–308 (2009)
5. Kawai, Y., Takahashi, K., Nagata, S.: PsyKolo3D-Interactive Computer Graphical Content of kolam design blocks. Forma **22**(1), 113–118 (2007)
6. Nagarajan, V.: Feeding a Thousand Souls: Women, Ritual, and Ecology in India - An Exploration of the Kolam. Oxford University Press, New York (2019)
7. Nagata, S., Robinson, T.: Digitalization of Kolam Patterns and Tactile Kolam Tools, Formal Models, Languages and Applications, Series in Machine Perception and Artificial Intelligence 66, World Scientific, pp. 353–362 (2006)
8. Nagata, S.: Loop Patterns in Japan and Asia. Forma **30**, 19–33 (2015)
9. Narasimhan, R.: The Oral-literate Dimension in Indian Culture, Indological Essays. In: Lockwood, M. (ed.) Commemorative Volume II for Gift Siromoney, Madras Christian College, pp. 67–79 (1992)
10. Prusinkiewicz, P., Krithivasan, K., Vijayanarayana, MG.: Application of L-systems to Algorithmic Generation of South Indian Folk-art Patterns and Karnatic Music. In: Narasimhan, R. (ed.) Commemorative Volume for Gift Siromoney. Series in Computer Science, vol. 16, pp. 229–247. World Scientific (1989)

11. Robinson, T.: Extended pasting scheme for kolam pattern generation. Forma **22**(1), 47–54 (2007)
12. Robinson, T.: The progressive step method for teaching geometry constructions to students with visual impairments. Japan J. Special Educ. **44**(6), 543–550 (2007)
13. Rosenfeld, A.: Picture Languages. Academic Press, London (1979)
14. Siromoney, G.: South Indian Kolam Patterns, Kalakshetra Q. **1**(I), 9–14 (1978)
15. Siromoney, G. Chandrasekaran, R.: On Understanding Certain Kolam Designs, Second International Conference on Advances in Pattern Recognition and Digital Technique, 6–9 January at the Indian Statistical Institute, Calcutta (1986). (https://www.cmi.ac.in/gift/Kolam/kola_understanding.htm)
16. Siromoney, G., Siromoney, R., Robinson, T.: Kambi kolam and cycle grammars, In: Narasimhan, R. (ed.) A Prespective in Theoretical Computer Science, Commemorative Volume for Gift Siromoney. Series in Computer Science, vol. 16, pp. 267–300. World Scientific (1989)
17. Smith, D.W., Smothers, S.M.: The role and characteristics of tactile graphics in secondary mathematics and science textbooks in Braille. J. Vis. Impairment Blindness, AFB **106**(9), 43–554 (2012)
18. Subramanian, K.G., Saravanan, R., Robinson, T.: P Systems for Array Generation and Application to Kolam Patterns. Forma **22**(1), 47–54 (2007)
19. Yanagisawa, K., Nagata, S.: Fundamental study on design system of kolam pattern. Forma **22**, 31–46 (2007)
20. Waring, T.M.: Sequential encoding of tamil kolam patterns. Forma **27**(1), 83–92 (2012)

Teaching Advanced Concepts Using Tangible Machines

QuBobs Teaching Kits to Explain Quantum Computing

Sophie Laplante[1(✉)], Loris Perez[2], Sylvie Tissot[3], and Lou Vettier[4]

[1] IRIF, Université Paris Cité, Paris, France
laplante@irif.fr
[2] Paris, France
[3] Anabole, Paris, France
[4] Lou Vettier Design, Paris, France
http://www.anabole.fr, http://louvettier.com

Abstract. We introduce a visual representation of qubits to assist in explaining quantum computing to students who lack the formal mathematics, and to a broad audience. The representation follows from physical devices that we developed to explain superposition, entanglement, measurement, phases, interference, and quantum gates. We have developed several hands-on teaching kits for students to manipulate objects that use this representation and to explain some applications, including quantum teleportation, quantum cryptography, and quantum nonlocality.

1 Introduction

This project is the result of a collaboration between a theoretical computer scientist, a designer, a software developer and a music composer. Together, we are developing tools to make quantum computing accessible to a wider audience, without sacrificing (too much) mathematical correctness. In this paper, we describe the mathematics behind the objects that we are developing, and provide a high-level view of the teaching kits we have developed based on this notation. We briefly and informally describe the pedagogical and design principles that have guided our work. Our representation can be used for teaching at the undergraduate and graduate level, to students with or without mathematical background, as well as for outreach activities. It evolved out of years of teaching quantum computing and other mathematics based theory of computing and algorithms. This is a practical report describing the materials developed and the underlying mathematics.

Most researchers in quantum information and algorithms agree that explaining quantum computing to a wide audience without relying on the language of mathematics is a difficult task. An article of Aaronson lays out the difficulties particularly well [Aar21]. Many different approaches have been used to explain quantum computing to a wide audience, and although there are far too many to give an exhaustive survey here, we give a few examples.

L. Perez—Musician and Composer.

© The Author(s), under exclusive license to Springer Nature Switzerland AG 2025
H. Fernau et al. (Eds.): CMSC 2024, LNCS 15229, pp. 77–92, 2025.
https://doi.org/10.1007/978-3-031-73257-7_6

For single qubit applications such as key agreement, Charles Bennett [Ben21] and many others use double arrows to represent photon polarization. John Preskill [Pre16] uses colored balls to represent 0 and 1. Qubits are represented as boxes and gates with two doors. Putting in a qubit in the top door and taking it out of the bottom door applies a change of basis. This representation is used for single qubit protocols as well as to explain entanglement.

Karl Svovil explains quantum cryptography using chocolate balls [Svo06] with a 0 or 1 symbol on each ball, in green or in red to indicate the basis. Special glasses (red and green) are used to explain measurement in the correct basis.

The most complete representation we are aware of is proposed by Economou, Rudolph and coauthors [Rud17, ERB05, EB22]. Qubits are presented as clouds of black and white balls, in a proportion that corresponds to the probability of getting the corresponding outcome when measuring the qubit. Blocks represent operations on qubits. Using this representation, they explain several advanced applications, such as Grover search [Gro96] and CHSH games [CHSH69].

For more advanced topics in quantum information, tensor networks have been used as a graphical language [WBC15, Bia19] to represent quantum states and processes.

Our approach has been to construct devices to help students develop a concrete mental image of the concepts underlying quantum computation, such as randomness and entanglement. Once the physical representation is clear, we move on to a slightly more abstract graphical representation, on which we can apply various quantum operations.

2 Representing Single Qubits

Just as the binary digit, or bit, is the basic unit of information in classical computing devices, the qubit is the basic unit of quantum information. A qubit is best described by comparison to a random bit, which is 0 with some probability $p \geq 0$ and 1 with probability $1-p$. A random bit can be thought of as the result of tossing a biased coin, prior to observing the outcome. Mathematically, one would associate a biased random bit with a two-dimensional vector with real coordinates p and $1-p$. A quantum binary state can be in a *superposition* of two basis states, for instance, an ion can be in a fundamental state (0) or an excited state (1), or it can be both at once, with varying proportions. A qubit is similar to a random bit, but instead of a probability, a complex coefficient, or amplitude, is associated with each outcome. Mathematically a qubit is represented by a two-dimensional complex vector (α, β) of L_2 norm 1, i.e. $|\alpha|^2 + |\beta|^2 = 1$, where $|\cdot|$ is the complex modulus or magnitude.

When a quantum binary state is measured, one of the outcomes will occur, with a probability given by the square of the modulus of the corresponding amplitude. After measurement, the superposition is said to collapse to the corresponding basis state, that is, the state becomes the qubit with amplitude 1 on the observed basis state and amplitude 0 on the other basis state.

2.1 Single Qubits

Our main contribution is a simple, accurate and surprisingly versatile representation of qubits. In Dirac's "bra-ket" notation, $|0\rangle$ ("ket zero") is the unit column vector $(1,0)^T$ and $|1\rangle$ is the column vector $(0,1)^T$. These form an orthonormal basis and linear combinations are written $\alpha|0\rangle + \beta|1\rangle$, where the coefficients α and β are called amplitudes. We represent the magnitude of each amplitude by a slice of a two-color disc. Blue corresponds to the magnitude of α and orange the magnitude of β. The proportion of each color represents the probability of measuring the corresponding outcome, hence (according to quantum mechanics) the blue part takes up an $|\alpha|^2$ fraction of the disc's area (and circumference) and the orange part takes up $|\beta|^2$ (Fig. 2).

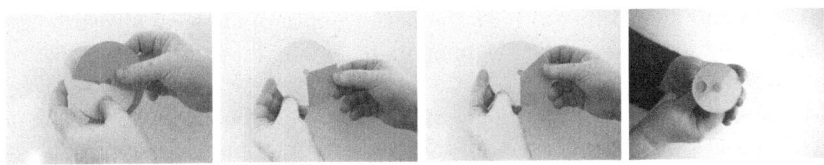

Fig. 1. An object that represents a quantum state $|\phi\rangle$. The squared magnitude of $\langle\phi|0\rangle$ is blue, and the squared magnitude of $\langle\phi|1\rangle$ is orange. On the device on the right, the window spins around the edge of the qubit. When it stops, the color under the window represents the outcome of the measurement. (Color figure online)

We illustrate the measurement of a qubit with a small device that spins around the qubit (Fig. 1). When it stops, a small round window reveals one of the colors. This is the outcome of the measurement. The probability of each outcome is the part of the disc that is the corresponding color, which is exactly the probability of measuring the qubit in the standard basis.

When an amplitude α is complex, our representation uses only its magnitude, and the phase is discarded (at least in the early phases of our presentations). Thankfully, most elementary quantum algorithms and use signed reals as amplitudes. (This is without loss of generality as a second qubit can be used to represent the phase.) To complete the description of the quantum state, we use a negative sign to indicate when the phase is negative. Sacrificing the phase is a useful first step towards understanding the inherent randomness of quantum measurements. In our presentations, we show how flipping a coin and observing the outcome is similar to measuring a qubit, but as we point out, a coin flip is not a quantum phenomenon. We then highlight the fact that the phase is a crucial part of a quantum state, and that without phases, we are only describing probabilistic binary states, limiting us to probabilistic computation.

2.2 Phases and Interference

Understanding phases and quantum interference is key to understanding the difference between classical states (or bits) and quantum states (or qubits). Amplitudes are complex numbers, which is bound to cause alarm in all but the most mathematically sophisticated audiences. Complex numbers are a convenient all-in-one notation that can be unpacked as a magnitude and a phase. We prefer to think of amplitudes as sine waves, with a magnitude representing the height of the wave and a phase representing a shift or a delay; a negative amplitude is just the inversion of the sine wave. Note that this is precisely where the so-called "wave-particle duality" of quantum mechanics manifests itself: quantum states behave as particles with a probabilistic component (determined by the magnitude) and as waves (with a phase).

The audiences we have encountered are familiar with how noise-canceling headphones work: sample the ambient noise, invert the wave, and add it to the audio input, and the noise present in the input is cancelled out. This is called interference, and it is precisely the same phenomenon that can be harnessed to get quantum algorithms to beat the performance of classical algorithms. We explain that a negative amplitude amounts to having flipped the wave-like component of the amplitude.

In simple quantum applications, such as key agreement or Deutsch's and Deutsch-Jozsa's algorithm, when interference occurs, opposing amplitudes cancel exactly. By this we mean that in the course of the computation, we obtain states of the form $\alpha|0\rangle - \alpha'|0\rangle + \beta|1\rangle + \beta'|1\rangle$ where $\alpha = \alpha'$. This is a great stroke of luck since $|\alpha - \alpha'|^2 = |\alpha|^2 - |\alpha'|^2 = 0$, and the slices with opposing phases cancel exactly. Obviously this does not hold in general. The fact that this happens to be correct in some key cases of interest allows us to convey how interference plays a role in quantum computation, in a meaningful and (almost) mathematically correct way, without fully explaining what happens in the general case. To more mathematically advanced audiences, we point out that we are taking shortcut. In a classroom situation we can return to the more conventional mathematical representation, and address other instances of interference, when discussing Grover's algorithm, or quantum strategies for Bell inequality violations for instance.

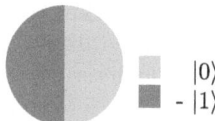

Fig. 2. The qubit $\frac{1}{\sqrt{2}}(|0\rangle - |1\rangle)$ represented as $(1/2, -1/2)$. (Color figure online)

3 QuBobs Representation of Two-Qubit Systems

In this section we present the graphical representation that we use to represent two-qubit systems with real amplitudes.

For the purposes of this paper, we write $|\phi\rangle \mapsto (P, Q)$ to mean that (P, Q) is our representation of $|\phi\rangle$. Using this notation, a qubit $\alpha|0\rangle + \beta|1\rangle$ with $\alpha, \beta \in \mathbb{R}$

is represented by $(sgn(\alpha)|\alpha|^2, sgn(\beta)|\beta|^2)$, and conversely, (A, B) is a representation of the qubit $sgn(A)\sqrt{|A|}|0\rangle + sgn(B)\sqrt{|B|}|1\rangle$. When several qubits are involved, since qubits can be entangled, the angle of the disk becomes important, and we will assume that, unless indicated otherwise, the blue part starts at the top going clockwise.

3.1 Two Qubits

The surprising expressiveness of our representation of qubits appears when we consider two qubits.[1] Separable states can be represented by two observation windows spinning independently over two qubits. To represent entanglement, we link the two spinning windows with a cog so that they spin synchronously (Fig. 3).

Fig. 3. Two ways of representing two entangled qubits. On the left, a cog makes both windows spin synchronously. On the right, the qubits are stacked and a single window lets us observe both qubits. (Color figure online)

Dirac's bra and ket notation extends to representing two qubits. A single qubit is a two-dimensional object. A two-qubit system is a four-dimensional object, with four amplitudes for each of the basis elements which we equate with boolean strings of length two. An arbitrary two-qubit (pure) state is written as $|\phi\rangle = \alpha|00\rangle + \beta|01\rangle + \gamma|11\rangle + \delta|10\rangle$. If we measure both qubits, then the outcome will be $00, 01, 11$ or 10 with probability $|\alpha|^2, |\beta|^2, |\gamma|^2$ and $|\delta|^2$, respectively.

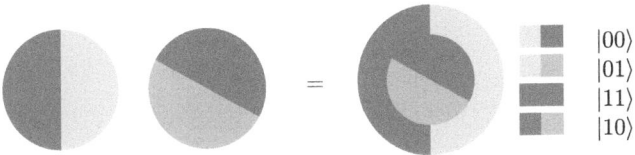

Fig. 4. Two representations of the same pair of entangled qubits, $\alpha|00\rangle + \beta|01\rangle + \gamma|11\rangle + \delta|10\rangle = \frac{1}{\sqrt{3}}|00\rangle + \frac{1}{\sqrt{6}}|01\rangle + \frac{1}{\sqrt{3}}|11\rangle + \frac{1}{\sqrt{6}}|10\rangle$. On the left they are represented side by side like in the device with two windows spinning synchronously. On the right the qubits are stacked the same way as in the device with a single window. (Color figure online)

---

[1] We use two color schemes, one for each qubit. The first qubit is light blue and bright orange; the second qubit is dark blue and light orange.

The key observation is that for any $\alpha, \beta, \gamma, \delta$ (properly normalized), when the observation windows are lined up (say they both start from the top of the disk and turn synchronously) it is always possible to line up the two discs so that the probability of getting 00 when both qubits are measured is $|\alpha|^2$, and similarly for 01, 11, and 10. To do this, line up the first qubit so that it is blue starting from the top going clockwise, and line up the second qubit so that it is orange starting at an angle $\theta = |\alpha|^2 \cdot 2\pi$ from the top, going clockwise. Then starting from the top and going clockwise, the two qubits are blue for $|\alpha|^2 \cdot 2\pi$, then the first qubit is blue and the second is orange for $|\beta|^2 \cdot 2\pi$ and so on.

Using two qubits side by side, it is easy to visualize the qubits separately, but the correlations are easier to visualise when we stack the qubits. We use a second device to visualize the four areas corresponding to each of the outcome pairs 00, 01, 11, and 10 (Fig. 4, on the right). With the two qubits stacked on top of each other, with one slightly smaller than the other, we can use a single window going around the perimeter. Going back and forth between these two representations gives us two complementary views of entanglement. The stacked representation also makes it clear that two qubits are a four-dimensional system.

When explaining protocols or algorithms, we set aside the physical devices with the windows spinning around the perimeter, and work with the diagrams illustrated in Fig. 4.

3.2 Separability and Entanglement

Two qubits are called separable if they can be written as the tensor product of two individual qubits. If they are not separable, they are said to be entangled. Separable states can also be represented with synchronized observation windows. For example, the state $\frac{1}{\sqrt{2}}(|0\rangle + |1\rangle) \otimes \frac{1}{\sqrt{2}}(|0\rangle + |1\rangle)$ can be represented with two half blue, half orange disks placed perpendicularly (Fig. 5).

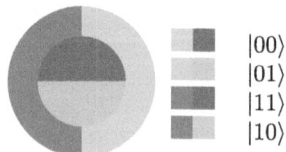

Fig. 5. The separable state $\frac{1}{2}(|00\rangle + |10\rangle + |11\rangle + |01\rangle) = \frac{1}{\sqrt{2}}(|0\rangle + |1\rangle) \otimes \frac{1}{\sqrt{2}}(|0\rangle + |1\rangle)$ (Color figure online)

This representation works well for two qubits. From three qubits onward we can no longer represent arbitrary states with discs with two colors arranged into two contiguous colored slices. However, it turns out that for many applications, two qubits suffice. The property of having qubits presented as two contiguous slices is also lost as a result of applying quantum gates (see Appendix A.1).

3.3 Partial Measurements

A measurement of one qubits in a two-qubit state is called a partial measurement. The state that remains after a measurement is made (called the residual state) is explained formally using the projection operator. Our representation is surprisingly successful in conveying the math without writing out any equations.

When we measure the first qubit of the state $|\phi\rangle = \alpha|00\rangle + \beta|01\rangle + \gamma|11\rangle + \delta|10\rangle$, the blue and orange parts of the first qubit split the second qubits into two areas: the one where the first qubit is blue, and the one where the first qubit is orange. If the first qubit is measured with a standard measurement, the outcome is 0 with probability $|\alpha|^2 + |\beta|^2$ and 1 with probability $|\gamma|^2 + |\delta|^2$. Similarly, if the second qubit is measured, the outcome is 0 with probability $|\alpha|^2 + |\delta|^2$ and 1 with probability $|\beta|^2 + |\gamma|^2$. In Fig. 6 we give an example of how we can visualize the residual state of the second qubit given that the first qubit has been measured. If the first qubit came out blue, then the observation window could have fallen in any place where the first qubit was blue. This part of the disk, on both qubits, since the measurement windows are linked by a cog, is the residual state. The rest of the disks are no longer accessible. Having this visual representation in mind makes it easier for students who move on to the formal mathematics notation to conceptualize what the projection does, why renormalization is necessary, and what the norm of the residual state (which is used to renormalize) represents.

Fig. 6. Visualizing the effect of a partial measurement. On the left, the qubit before measuring the fist (outer) qubit. On the right, the residual state when the outcome is blue (middle) or orange (on the right). (Color figure online)

4 Hands-On Learning Kits

We have developed five hands-on teaching kits: qubits and entanglement (Sect. 4.1); one-qubit circuits (Sect. 4.2); key agreement (BB84) (Sect. 4.3); teleportation (Sect. 4.4); and quantum non-locality (Sect. 4.5). We have used the first four kits with high school students and members of the general public. Going through one kit takes around one hour, more or less depending on the audience. Although we have not caried out formal evaluations of our approach, audience response has been very positive so far, particularly with high school students who are interested in science, as well as adults who have already expressed interest in quantum computing only to be left dangling when it comes to truly understanding where the quantum advantage comes from. We provide a brief overview of each of these kits in the next sections.

4.1 Qubits and Entanglement

Fig. 7. Our Entanglement kit allows students to manipulate 2-qubit systems and become familiar with entanglement. It uses three different representations, two which are visual and one which makes the connection to linear algebra notation. (Color figure online)

This kit contains a game board with two individual "qubit" devices, and an additional device that represents the same two qubits (Fig. 7). The three paper devices can be manipulated to obtain the desired amplitudes. A stack of game cards contains challenges, ordered by difficulty, with the answer on the reverse. The challenges are to complete the board given a description of the two-qubit system, either as the two individual states in the graphical representation, or as the amplitudes of the system, or some combinations thereof. The amplitudes are given in column (vector) form so as to gently lead into the math notation, although we intentionally do this without calling attention to this fact.

One of the features of this board is that we can manipulate the 2-qubit system as two 2-dimensional objects (that is, these objects have two moving pieces), together with their correlation, or as a 4-dimensional object (similarly, the corresponding object has 4 moving pieces). It provides a very intuitive understanding of the partial trace operation (deriving one of the individual qubits from a two-qubit system), and is a great visual aid in understanding partial measurements (when one of the qubits is measured, the residual state on the second qubit is changed, the quantum analogue of taking a maginal distribution in a bipartite distribution). This is key in understanding quantum algorithms, and in particular this is key for understanding the teleportation protocol (Sect. 4.4).

4.2 One-Qubit Circuits

Quantum computation consists in applying a sequence of few-qubit operations to an initial state, and at the end of the computation, making a measurement to obtain the end result of the computation. In order to familiarize students with basic quantum operations (X, the negation gate; Z, the phase-flip gate; and H, the so-called Hadamard gate) we developed a 1-qubit circuit kit. We use the same notation for circuits as are used in the literature and textbooks, with computation going from left to right, starting from an initial state (usually the all-blue qubit). Students moving on to more advanced topics will be already familiar with the standard circuit notation. Game cards are provided that have challenges of increasing difficulty, with solutions on the reverse side. Either the operations are given and the results should be found, or the results are given and the operations should be found. Operations are given as small cards and printed on the back of the card is a cheat-sheet giving the effect of the operation (Fig. 8).

Fig. 8. Manipulating 1-qubit circuits. (Color figure online)

4.3 Key Agreement

Our kit to explain quantum cryptography and the Bennett and Brassard "BB84" key agreement protocol [BB84] contains Qubits with $|0\rangle$ (blue) on one side and $H|0\rangle$ (the Hadamard operation applied to the state 0) on the other; qubits with $|1\rangle$ (orange) on one side and $H|1\rangle$ on the other; two dice, one to (tentatively) pick a bit of the key, and one to apply, or not, the Hadamard operation to the bit; eight numbered black envelopes in which qubits can be placed and observed on either side using small cutout flaps; a grid to mark the results: what value was selected, whether Hadamard was applied, what measurement was made, and what the outcome was (Fig. 9).

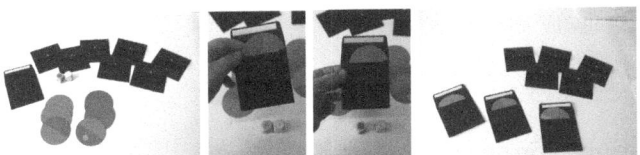

Fig. 9. Our BB84 kit comes with eight envelopes and two dice. Alice picks 8 qubits and places each them in a numbered envelope. Qubits can be measured by opening a small flap in the envelope. Flipping a qubit over changes the basis. Flipping over the envelope changes the measurement basis. (Color figure online)

For readers familiar with the BB84 protocol, we briefly describe how the kit is used. Participants playing Alice can prepare qubits to be sent to Bob by placing them in envelopes with the correct side facing up. A second participant playing Bob can measure a qubit in the standard basis by opening a flap on the front of the envelope, or in the Hadamard basis by turning over the envelope (and hence the qubit inside) and opening a flap on the back of the envelope. All the results are marked on a grid. Participants can see for themselves that any time the basis chosen for the qubit and the basis for the measurement coincide, the outcome of the measurement made by Bob equals the bit that was chosen by Alice. We can then explain that any discrepancy would be due to an observation made by a third party.

4.4 Teleportation

One of our more advanced kits covers quantum teleportation, invented by Bennett et al. [BBC+93]. Teleportation is a two-player protocol that allows Alice to transmit an arbitrary qubit to Bob, with the help of a uniform entangled pair of qubits that they have shared prior to the start of the teleportation protocol, plus two bits of classical communication. In this protocol, entanglement is used by Alice to correlate her qubit with the shared entangled pair, and after measuring her part of the state, she sends two classical messages to Bob to correct the residual state on his side, so that he is left with Alice's qubit at the end of the protocol.

Our teleportation teaching kit allows students to walk through the teleportation protocol, and to verify that the qubit is correctly transmitted in all of possible executions (Fig. 10).

Fig. 10. Our teleportation kits lets students walk through the teleportation circuit. (Color figure online)

4.5 Quantum Non-locality

Our most ambitious kit concern quantum non-locality, a phenomenon that Einstein called "spooky action at a distance" [EPR35]. Quantum non-locality occurs when two parties carry out measurements on a shared bipartite system, and the probability distribution that they obtain cannot be obtained by classical means. In the usual scenario, Alice has two possible inputs, or measurement settings, and Bob also has two possible inputs. Depending on their input, they can make any local operation on their part of the system, and then make a measurement. (This framework was defined by John Bell [Bel64]).

These games have been used experimentally to demontrate the validity of quantum mechanics. If one can implement such a game with quantum entanglement, and establish that the resulting behavior cannot be explained by shared randomness alone, then this establishes that quantum mechanics is necessary to explain the results of the experiment. A large number of experiments have been carried out to this end, starting in the 1970s with Clauser and Freedman [FC72] and all "loopholes" being closed in 2015 [HBD+15].

Most experiments are based on the proposal of Clauser Horne Shimony and Holt [CHSH69]. Our kit presents it as a card game. Each player is dealt a card, and based on the card should make some measurement. If both players get card number 1 then they win if their outcomes are different. In all other cases their outcomes should be the same. The kit contains a two-sided token used to pick an outcome. The players convince themselves that there is no classical strategy to win this game more than 75% of the time. The second part of the kit walks them through a quantum strategy that wins with probability around 85%.

The key difficulties in making the quantum strategy accessible were (1) simplifying the quantum protocol so that it is as easy as possible to see what the winning probabilities are, and (2) simplifying the calculations that lead to the correct winning probability. In the standard presentation, the winning probability is $\cos^2(\pi/8)$, and computing this probability requires some trigonometry, in addition to linear algebra. We use 6/7 as an approximation. A second obstacle is computing the result of interference. We reduce the calculation to two cases, which are illustrated in a cheat sheet which is provided with the cards.

5 Design Principles and Pedagogical Considerations

Throughout the process of designing objects and hands-on teaching kits we have been guided by a handful of design principles and pedagogical considerations. We provide a brief and very informal description here.

Our pedagogical approach has been guided by our experience in teaching theoretical subjects to Computer Science undergraduates, where we have found it very useful to use simple, everyday objects to ground mathematical thinking before moving on to formal mathematical notation. We believe that mathematical formalism can be one of the biggest obstacles to approaching mathematical topics. Mathematics seems to elicit a mixture of fear, a feeling of inadequacy, and leads to a loss of self confidence in all but the most mathematically sophisticated audiences. Using familiar objects and appealing to people's natural abilities (including their natural spatial abilities [Gia07]) we hope to alleviate the fear and restore self confidence.

Many mathematicians rely heavily on diagrams and drawings to manipulate mathematical constructs, and mathematical notation is used not so much as a thinking tool, but as a way to formalize and communicate mathematical arguments to other mathematicians [Had54]. It seems only natural that we should be able to use visual representations to convey mathematical constructs.

We want the objects we use to draw people in, and spark curiosity and an eagerness to learn. However, we take care that the objects themselves do not overshadow the learning experience so we avoid anything too catchy. We design the objects so that they are easy to use and pleasant to touch. We use paper whenever possible, because it is familiar and non-intimidating. Our devices are meant to be played and experimented with, with the intent that this should help anchor thought processes in something concrete instead of abstract. Our boards are designed to be used with dry-erase markers so participants can use trial and error to explore various possibilities.

Our online applications use the same visuals as the paper objects, and we design the interface so that it feels similar to manipulating the physical objects. Music is used in our digital applications as well as videos. Although the music is designed to match somewhat rigorously our visual representation of qubits, our hope is that this would not be so overtly apparent as to distract viewers from the concepts, but instead that it contribute to a more general sensory experience and, again, a feeling of familiarity.

A huge effort goes into presenting things in the simplest way possible, starting with the underlying mathematics, but also in the design of our objects. Our kits are reminiscent in form of board games. In designing them, we make sure the pieces are easy to manipulate. This means we try to have as few pieces as possible so that it is easy to find the right piece and to put in the right place. Any fiddly bits would add to a sense of frustration and take attention away from the task at hand. Pieces are often two-sided so that there are basically two options to choose from. Game pieces include dice for randomness, playing cards for inputs, and tokens, again in an attempt to instill a sense of familiarity.

All of these are important in our process but mathematical accuracy always comes first. Because everything is designed with the underlying mathematics in mind, teachers can then introduce the formal mathematical notation to more advanced students, which then allows students to build upon what they have understood and move onto more advanced material.

Finally, we always strive for inclusiveness. The choice of colors and shades are selected to be distinguishable by the color blind. We plan to use music and sound to make some of our tools accessible to the visually impaired. We avoid any imagery that is associated with stereotypes about (computer) science and (computer) scientists, or with gender biases, and we try to make sure any photos we use on the web site and social media represent diversity in all its forms. We do this so that everyone can feel like we are addressing them, and not some specific age group, gender, educational level, or any other category they might not feel a part of. If our goal is to make quantum computing accessible, it has to be accessible to all.

6 Conclusions

We have presented new physical devices and a visual representation of qubits that can be used for teaching or for outreach activities. We have found it particularly helpful in conveying concepts such as entanglement and interference, concepts that key to understanding quantum algorithms, but can be difficult to grasp at an intuitive level, especially for audiences who are not accustomed to using mathematical formalism.

More information on our project can be found on the website https://qubobs.irif.fr.

Acknowledgements. We would like to thank the students and participants in our outreach activities for their feedback. We thank Maris Ozols for references to tensor diagrams and Serge Massar for references to some of the other representations. We are grateful to the many colleagues who served as sounding boards throughout the different versions of this work.

This project was funded by IRIF, by a grant from IDEX Université Paris Cité, and the ANR grant SAPS-RA-MCS QUBOBS.

Appendix

A Quantum Operations

A.1 Single Qubit Gates and Linearity

Just as a classical computation can be viewed as applying logical operations to bits, a quantum computation consists of applying quantum operations, or quantum gates, to qubits. Quantum gates are unitary maps (they are linear, invertible, and norm-preserving).

We illustrate quantum gates by flipping over a disk to reveal the effect of a gate on each of the basis states.[2]

In order to explain simple quantum circuits, we apply quantum gates on our representation of qubits by appealing to linearity. One has to proceed with extreme caution since the linear maps are defined over the amplitude vectors and our representation operates on the moduli squared of the amplitudes.

Let us consider the case of the Hadamard gate, one of the most fundamental gates in quantum computing. Hadamard applied to blue is the state composed of half blue and half orange, which we denote by $(1/2, 1/2)$. Similarly Hadamard applied to orange is $(1/2, -1/2)$. Notice that this introduces a negative phase. If we apply Hadamard on $(1/2, 1/2)$, we apply it to both parts. The blue half becomes $(1/2, 1/2)$ and the orange half becomes $(1/2, -1/2)$, so we obtain four parts, which we will write $(1/4, 1/4, 1/4, -1/4)$. Up to this stage, our representation accurately reflects the effect of the gate. What can get us into trouble here is how we handle the addition of the two blue parts, and the addition of the two orange parts, one with an opposite signe. This is what is called positive and negative interference. We are tempted to say that the orange parts with opposite signs cancel, leaving an all-blue disk. In this case, this correctly mimics what happens to the amplitudes, and it allows us to convey a message that is essentially correct: the fact that applying Hadamard twice to basis states is identity, and this occurs because of interference. It is also all that is needed to explain how and why some simple applications, such as the BB84 key exchange protocol, or Deutsch's algorithm, work.

---

[2] Many of the gates we consider are self-adjoint, so flipping over twice gets us back to the initial state.

What we are sweeping under the carpet here is that when we add the blue parts, the actual calculation is over the amplitudes: $\sqrt{1/4} + \sqrt{1/4}$ (which is 1) as opposed to $1/4 + 1/4$ (which is $1/2$). The fact that Hadamard is unitary is what guarantees that when the orange parts cancel out (as they do here since $\sqrt{1/4} - \sqrt{1/4} = 0$) the blue parts add up to 1. A more detailed explanation of interference is required to explain some algorithms and experiments (such as the CHSH game) but we leave a detailed discussion of how to explain quantum interference to a subsequent paper.

Returning to quantum gates, we consider the effect of single qubit gates G, with $G|0\rangle = a|0\rangle + b|1\rangle$ and $G|1\rangle = c|0\rangle + d|1\rangle$ We will say that in our representation, $G(P,Q) = (A, B, C, D)$ where

$$A = P \cdot \text{sgn}(a)|a|^2 \quad B = P \cdot \text{sgn}(b)|b|^2$$
$$C = Q \cdot \text{sgn}(c)|c|^2 \quad D = Q \cdot \text{sgn}(d)|d|^2$$

In the schematic representation, we obtain four parts, starting from the top going clockwise with blue, and alternating blue and orange parts (Fig. 11).

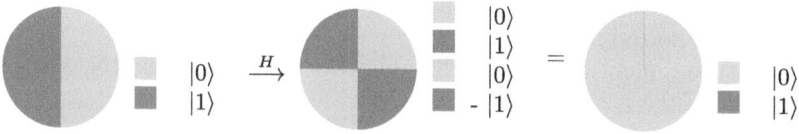

Fig. 11. Applying Hadamard to $\frac{1}{\sqrt{2}}(|0\rangle + |1\rangle)$ behaves well since the positive and negative parts cancel completely. (Color figure online)

We use the notation (A, B, C, D) to describe a disc with four areas, whose colors are blue, orange, blue, orange. For example, if Hadamard is applied to $(1/2, 1/2)$, we get $(1/4, 1/4, 1/4, -1/4)$. Unless the state is entangled with another, we start from the top with a blue slice of size A, and so forth, going clockwise (Fig. 12).

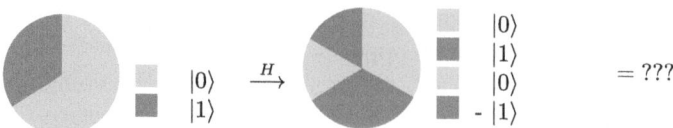

Fig. 12. Hadamard applied to other states requires some additional reasoning in order to simplify the state. Applying Hadamard to $\frac{\sqrt{2}}{\sqrt{3}}|0\rangle + \frac{1}{\sqrt{3}}|1\rangle$, we get $\frac{1}{\sqrt{3}}|0\rangle + \frac{1}{\sqrt{3}}|1\rangle + \frac{1}{\sqrt{6}}|0\rangle - \frac{1}{\sqrt{6}}|1\rangle$. Our representation might suggest that there should be $\frac{1}{3} - \frac{1}{6}$ orange (1) left, whereas the correct fraction is $(\frac{1}{\sqrt{3}} - \frac{1}{\sqrt{6}})^2$. (Color figure online)

A.2 Applying Gates on two Qubits

Applying two-qubit gates works similarly to the one-qubit case. Recall that two qubits in our representation, (P, Q) and (P', Q'), do not fully determine the two-qubit state, since the angle at which they are placed determines how they are correlated. If the first qubit is blue going clockwise from the top, and the second disk is orange starting from an angle of $\theta \cdot 2\pi$, going clockwise, then (ignoring for the sake of simplicity any possible phases) the state represented is

$$|\psi\rangle = \sqrt{\theta}|00\rangle + \sqrt{P-\theta}|01\rangle + \sqrt{1-P-P'+\theta}|11\rangle + \sqrt{P'-\theta}|10\rangle.$$

We will use the notation $|\psi\rangle \mapsto (P, Q)\theta(P', Q')$.

When we apply a single qubit gate to the first qubit of a two-qubit state, we apply it while preserving the angle between the two qubits. To ensure that entanglement is preserved, we apply it to all four areas corresponding to the basis elements in a two-qubit system, $|00\rangle, |01\rangle, |11\rangle, |10\rangle$. These four areas are easily seen in the stacked representation. Each of the four areas will be further subdivided by gates such as Hadamard. Similarly, when applying a two-qubit gate, we apply it to all four areas. Control gates are particularly easy to visualize. (See Fig. 13.) In the area where the control bit is 0, we do nothing, and in the area where the control bit is 1, we apply the gate as usual.

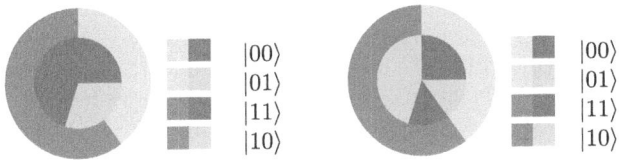

Fig. 13. The CNOT gate is applied to the qubit on the left. The control bit is the outer bit. When the outer bit is blue, the inner bit remains the same. When the outer bit is orange, the colors of the inner bit are flipped. (Color figure online)

References

[Aar21] Aaronson, S.: What makes quantum computing so hard to explain? Quanta Mag. (2021)

[BB84] Bennett, C.H., Brassard, G.: Quantum cryptography: public key distribution and coin tossing. Theoret. Comput. Sci. **560**, 7–11 (1984)

[BBC+93] Bennett, C.H., Brassard, G., Crépeau, C., Jozsa, R., Peres, A., Wootters, W.K.: Teleporting an unknown quantum state via dual classical and Einstein-Podolsky-Rosen channels. Phys. Rev. Lett. **70**(13), 1895–1899 (1993)

[Bel64] Bell, J.: On the Einstein-Podolsky-Rosen paradox. Physics **1**, 195–200 (1964)

[Ben21] Bennett, C.H.: Quantum information's revolutionary origins (2021). https://www.youtube.com/watch?v=B5BUhzBlO-U

[Bia19] Biamonte, J.: Lectures on quantum tensor networks. Technical report 1912.10049 (quant-ph), arXiv (2019)

[CHSH69] Clauser, J.F., Horne, M.A., Shimony, A., Holt, R.A.: Proposed experiment to test local hidden-variable theories. Phys. Rev. Lett. **23**(15), 880–884 (1969)

[EB22] Economou, S.E., Barnes, E.: Hello quantum world! A rigorous but accessible first-year university course in quantum information science. Technical report arXiv:2210.02868, ArXiv (2022)

[EPR35] Einstein, A., Podolsky, B., Rosen, N.: Can quantum-mechanical description of physical reality be considered complete? Phys. Rev. **47**(10), 777–780 (1935)

[ERB05] Economou, S.E., Rudolph, T., Barnes, E.: Teaching quantum information science to high-school and early undergraduate students. Technical report quant-ph 2005.07874, ArXiv (2005)

[FC72] Freedman, S.J., Clauser, J.F.: Experimental test of local hidden-variable theories. Phys. Rev. Lett. **28**(14), 938–941 (1972)

[Gia07] Giaquinto, M.: Visual Thinking in Mathematics. Oxford University Press, Oxford (2007)

[Gro96] Grover, L.K.: A fast quantum mechanical algorithm for database search. In: Proceedings, 28th Annual ACM Symposium on the Theory of Computing, pp. 212–219 (1996)

[Had54] Hadamard, J.: The Psychology of Invention in the Mathematical Field. Dover Publications (1954)

[HBD+15] Hensen, B., et al.: Loophole-free bell inequality violation using electron spins separated by 1.3 kilometres. Nature **526**(7575), 682–686 (2015)

[Pre16] Preskill, J.: Quantum computing and the entanglement frontier, 2016 Leigh Page Prize lecture series, hosted by Yale Department of Physics and Yale Quantum Institute (2016). https://www.youtube.com/watch?v=bPNlWTPLeqo

[Rud17] Rudolph, T.: Q is for Quantum. Self-published (2017)

[Svo06] Svovil, K.: Staging quantum cryptography with chocolate balls. Am. J. Phys. **74**(800) (2006)

[WBC15] Wood, C.J., Biamonte, J.D., Cory, D.G.: Tensor networks and graphical calculus for open quantum systems. Quantum Inf. Comput. **15**(9 and 10), 0759–0811 (2015)

Solid Geometry Modeling: 3D Printing Is Not Always the Best Option

Matthias Müller[1](✉) [iD], Benjamin Weißing[2], and Pascal Lütscher[1] [iD]

[1] University of Teacher Education of the Grisons, Scalärastrasse 17, 7000 Chur, Switzerland
matthias.mueller@phgr.ch
[2] Free University of Bozen-Bolzano, Piazza Università, 1, 39100 Bolzano, Italy

Abstract. The engagement with a fascinating geometric problem involving the creation of a three-dimensional solid model can be traced back to the 18th century: The goal is to find a solid that has a circular base, looks like a triangle from one side, and like a square from the other. This problem has inspired many mathematicians, such as Georg Pólya in 1966. The appeal of the problem has not diminished over the years, mathematicians and mathematics educators are still engaged with analog or slightly modified versions of the problem. As with historical examples, the problem suggests a unique solution, but many descriptions of solutions to the problem are incomplete, as there are infinitely many solids that meet the required properties.

This paper explores how to find different solutions that meet the specified conditions of the geometric problem. It will determine and compare the volumes of two solids that meet these conditions. Furthermore, the advantages of creating analog solid models, as were made in the 18th century, will be discussed in comparison to 3D printed models. This approach can be utilized in various learning environments, leading to an understanding of the problem according to modern problem solving theory in mathematics education.

Keywords: solid models · mathematical problem solving · 3D printing

1 A Mathematical Problem Drawn by History

The geometric problem to be studied can be traced back to the 18th century. Peter Friedrich Catel described the problem in his book *"Mathematisches und physikalisches Kunst-Cabinet"* from 1790. As a precision mechanic and toy manufacturer from Berlin, he crafted a board made of plum wood, which was equipped with three openings and is referred to as *"Die mathematischen Löcher"* (*Mathematical Holes*) [16]. The description of the problem is as follows [1]:

The mathematical holes […] consist of a board made of plum wood, 9 inches long and 2½ inches wide, containing a square, a round, and a triangular hole. The task is to specify the solid that can pass through all three holes and yet perfectly plug or fill them. Such solids can be cut from bread […] or wood.[1]

---
[1] Original phrasing: *"Die mathematischen Löcher […] bestehen aus einem von Pflaumenholz verfertigtem Brette, 9 Zoll lang und 2½ Zoll breit, worin ein viereckiges, ein rundes und ein*

A solution to this problem was addressed already in the aforementioned book. An illustration shows the board with the three openings and a suitable plug. However, the illustration in the book is not particularly informative and can only provide a preliminary idea of a possible solid with the required conditions. Nevertheless, the perspective drawing is remarkable for the capabilities of that time.

The Hungarian mathematician George Pólya revisited the problem 200 years later in the book *"Vom Lösen mathematischer Aufgaben"* in 1966. In it, the desired solid is referred to as a universal plug [14, p. 200]. The problem is formulated in a very similar manner, seeking a solid that has a circular base, looks like a triangle from one side and like a square from the other. The solid must fulfill three conditions [14, pp. 200]:

- (C1) The projection of the solid onto the plane is a circle (plan view).
- (C2) The projection of the solid onto the front is a square (elevation view).
- (C3) The projection of the solid onto the profile is a triangle (section view).

In this article we follow the description of the problem and we call the required solid *universal plug* such as other authors before [e.g. 13].

The allure of this problem has remained strong over the years, continuing to captivate mathematicians and mathematics educators who engage with its various versions, both similar and modified. Echoing historical instances, the problem hints at a unique solution, often referred to as the universal plug. This paper investigates the possibility of multiple solids satisfying the geometric problem's specified conditions. It aims to identify and compare the volumes of these qualifying solids. Additionally, it will introduce and evaluate methods for creating the universal plug, highlighting the benefits of analog models compared to digital techniques such as 3D printing.

2 Different Plugs with Different Volumes

To achieve the desired solid, the problem should first be precisely reformulated and the conditions relevant to its resolution clearly named. This formulation is abbreviated as P in the paper (Fig. 1).

> A workpiece has three openings. The side lengths of the square opening correspond to the diameter of the circular opening as well as the base of the triangular opening. The triangular opening is an isosceles triangle (see Fig. 1). The task is to find solids that can be inserted into the openings of the workpiece in a contour-equivalent manner.
>
> (P)

Contour-equivalent means that the solids completely fill the openings [6].

Regarding to problem statement P and according to the stated objective an examination of the uniqueness of the solution is a meaningful enterprise. Indeed, modern contributions to the topic suggest various solutions are possible [6]: There are many such solids. The simplest form, which also has the largest volume, is obtained by taking

---

dreieckiges Loch sind. Die Aufgabe davon ist: Dass man die Figur angeben soll, welche durch alle 3 Löcher gehen könne, und doch solche vollkommen verstopfe oder ausfülle. Man kann solche von Brot [...] oder Holz schneiden lassen." [1, p. 16].

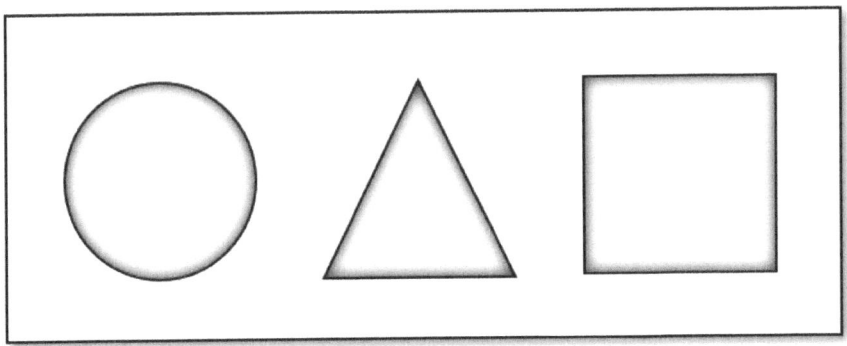

Fig. 1. Workpiece with three openings: circle, isosceles triangle, and square. The diameter of the circle corresponds to the base of the triangle as well as to one side of the square.

a straight cylinder whose height is equal to the diameter, and then removing two equal pieces through two flat cuts [6, p. 169].

Actually, volume is proposed as a structuring measure in various studies [e.g. 4, p. 58]. Initially, it makes sense to agree on the restriction that, according to the problem statement P, convex solids are sought. It is sensible to inquire about the solid with the maximum volume first. In the following, the volume of an universal plug of maximum volume will be determined.

2.1 Universal Plug of Maximum Volume

The convex solid which meets the conditions from P having maximum volume [4, p. 58] can be cut from a cylinder. The volume can be determined in various ways. We will explain an adjustable approach (see Fig. 2).

Cutting a cylinder parallel to the x-y-plane in the height l, $0 \leq l \leq h$, gives us the shape showed in Fig. 2: The upper semicircle can be described by $y = \sqrt{r^2 - x^2}$.

For $x_0 = x_0(l)$ we get:

$$\frac{x_0}{r} = \frac{h-l}{h} \quad (1)$$

$$x_0 = \left(1 - \frac{l}{h}\right) \cdot r \quad (2)$$

Calculating the area $A(l)$ of the shape in relation to l gives us:

$$A(l) = 2 \int_{-x_0}^{x_0} \sqrt{r^2 - x^2} \, dx \quad (3)$$

We get the maximum volume:

$$V_{max} = \int_0^h A(l) dl = \frac{3\pi - 4}{3} r^2 h \quad (4)$$

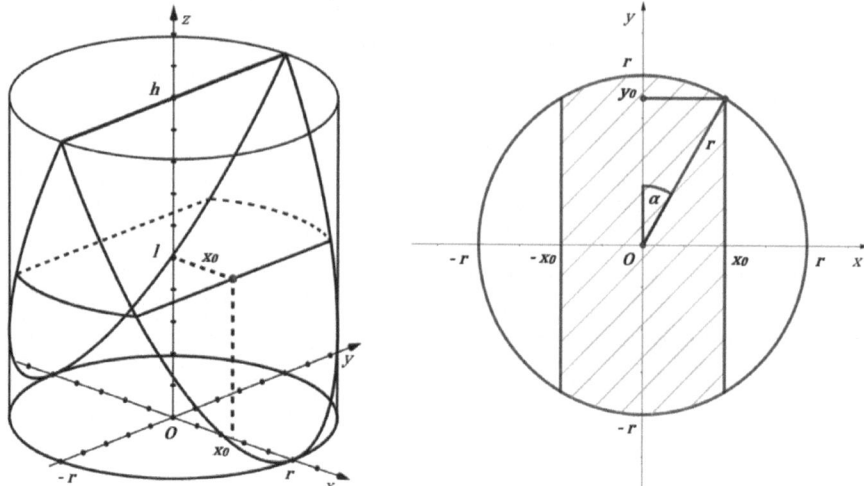

Fig. 2. Cutting a universal plug of maximum volume out of a cylinder: intersection solid (left drawing) and intersection area of height l (right drawing).

According to the conditions of P the height h of the solid equals the diameter of the cylinder. It is $2r = h$ with

$$V_{max} = \frac{3\pi - 4}{6} h^3 \qquad (5)$$

A comparison of the maximum volume with the volume of the cylinder leads to the following quotient:

$$\frac{V_{max}}{V_Z} = \frac{\frac{3\pi-4}{3} r^2 h}{\pi r^2 h} = \frac{3\pi - 4}{3\pi} \approx 0{,}575 \qquad (6)$$

The maximum volume is obviously larger than half the volume of the cylinder of the same height. A picture of the described required solid of maximum volume is given in Fig. 3.

2.2 Example of Non-Convex Universal Plug

A non-convex solid applying the conditions of P [4, p. 58] can be easily removed out of a cylinder, too.

In this case cutting the solid anywhere parallel to the x-y-plane gives us ellipses, which major axis are all the same length equal to the diameter of the cylinder. The minor axis of these ellipses is getting smaller with increasing height of cutting (see Fig. 4).

One option determining the volume of the required solid is to summarize all areas of these ellipses. Generally, we know this approach from the previous section:

$$V_{min} = \int_0^h A(l) dl \qquad (7)$$

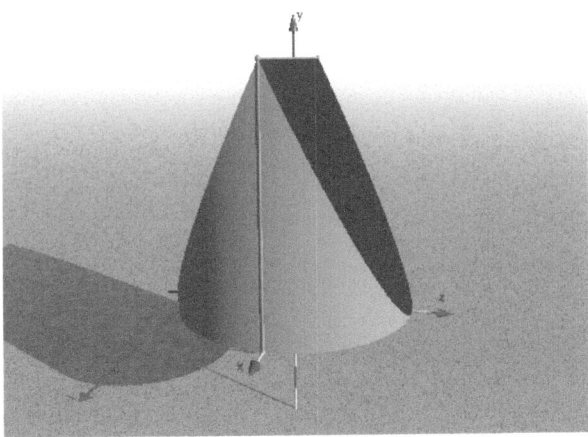

Fig. 3. Orthographic projection of a universal plug of maximum volume.

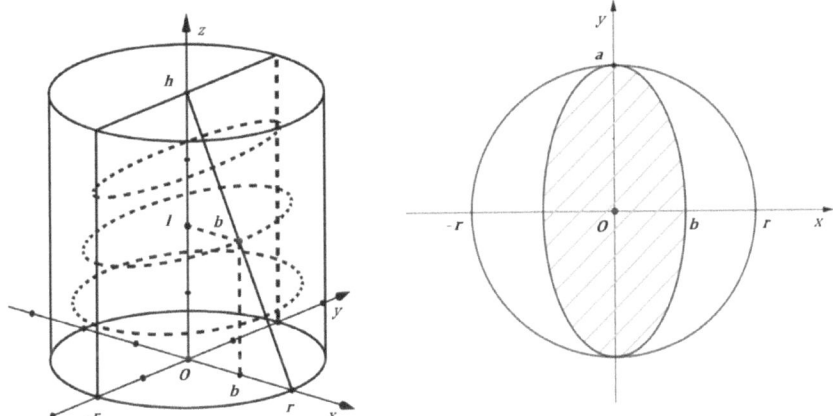

Fig. 4. Non-Convex Universal Plug: parallel ellipses in a cylinder (left drawing) and intersection area of height l (right drawing).

Cutting a cylinder parallel to *x-y*-plane in the height $l, 0 \leq l \leq h$, gives us the shape showed in Fig. 4. We see an ellipse with semi-major axis a and semi-minor axis b. Knowing the area of an ellipse in general:

$$A(a, b) = \pi ab \tag{8}$$

In our case the area is related to the variable semi-minor axis b. As described above we can set the semi-major axis a constant, and a equal to the radius r of the cylinder:

$$A(b) = \pi rb \tag{9}$$

The variable semi-axis $b = b(l)$ is given because of the Intercept Theorem (see Fig. 4):

$$\frac{b}{r} = \frac{h-l}{h} \tag{10}$$

$$b = \left(1 - \frac{l}{h}\right) \cdot r \tag{11}$$

We get the minimal volume:

$$V_{min} = \int_0^h \pi r^2 \left(1 - \frac{l}{h}\right) dl = \frac{\pi}{2} r^2 h \tag{12}$$

According to the conditions of P the height h of the solid equals the diameter of the cylinder. It is $2r = h$ with

$$V_{min} = \frac{\pi}{8} h^3 \tag{13}$$

A comparison of the maximum volume with the volume of the cylinder leads to the following quotient:

$$\frac{V_{min}}{V_Z} = \frac{\frac{\pi}{2} r^2 h}{\pi r^2 h} = \frac{\pi}{2\pi} = \frac{1}{2} \tag{14}$$

It can be noted; the volume of this non-convex universal plug equals half the volume of the given cylinder (Fig. 5).

Fig. 5. Orthographic projection of a non-convex universal plug.

3 Methods of Producing Universal Plugs

Different methods can be chosen to produce solids matching the conditions of problem P. Here, three methods are presented as examples.

An easy and empathic method suitable for creating a model of different universal plugs is the use of modeling clay. For learning environments for students, it is advisable using clay, which dries at room temperature. Moreover, it is advisable to design a framework in advance to ensure the model meets the known conditions. For this, figures of a circle, square, and isosceles triangle can be cut from cardboard. When these are assembled orthogonally to each other, they form a framework to which the modeling clay can be applied (see Fig. 6). The model can be smoothed after drying with sandpaper and files. This method in particular gives the impression that there can be more than one convex solid that meets the conditions of P, as shaping the models with modeling clay encourages various surface designs. Especially in combination with another method (e.g., 3D printing), a targeted comparison is possible.

Fig. 6. Framework of a universal plug made of cardboard (left) and a model of a universal plug made of modeling clay (right).

Of course, 3D printing can be one possible approach. Especially in STEAM learning environments it might be suitable. One handy software is *3D Builder* from Microsoft. The 3D models of typical solids such as cuboid and cylinder can be found by drag and drop. There is a feature of the program, which initiates the cutting solid. A 3D model of the universal plug of maximum volume could be designed easily in this way (see Fig. 7). A 3D model of a non-convex universal plug is much more difficult to design with the program. It is not a good option for a mathematical learning environment, if the intention is finding more then one solid matching the conditions of P.

To create a model of a non-convex universal plug, another method can be chosen. This involves creating many individual ellipses that are to be stacked on top of each other. This procedure resembles the process of determining the volume of the solid. (see Sect. 2.2). The ellipses can be plotted using dynamic geometry software (e.g. GeoGebra). The individual ellipses can be printed on cardboard. Naturally, this method

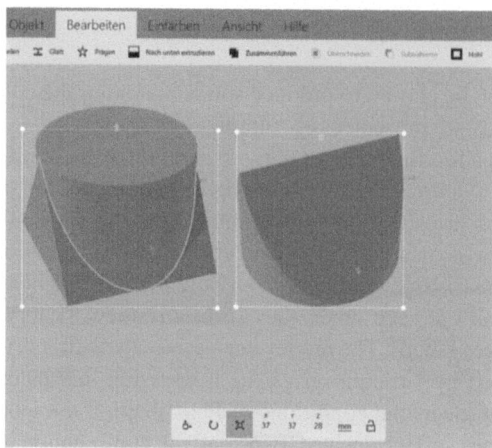

Fig. 7. Using Microsoft 3D Builder to design a universal plug of maximum volume.

produces a less precise model compared to, for example, 3D printing. However, the basic idea of layer-by-layer construction is comparable in both methods, which particularly illustrates the fundamental concept of determining volumes. In constructing the ellipses, it is helpful that the length of the major axis remains constant. The lengths of the minor axis can be calculated through a corresponding linear relationship. Once all lengths are calculated, the ellipses can be drawn and the cardboard can be punched, lasered, or cut. The production process in this method provides a tactile approach to determining the volume of the convex solid with minimal volume. The following Fig. 8 shows the individual (here lasered) ellipses and the assembled model of a non-convex universal plug.

> The advantages of the analog methods are
> (1) an illustration of the different approaches finding a solid which meets the given conditions,
> (2) an easy demonstration of the existence of infinitely many possible solids matching the given conditions, and
> (3) a visualization of the approaches of determining (and calculating) the volume of the different solids.

Dealing with Problem P in a mathematical learning environment using the described procedures is an example of a problem solving approach suitable for mathematics classroom as well as extracurricular after school courses. A determination of formulas of the volume of the different universal plugs such as described in Sect. 2 depends strongly on the age such as the mathematical competencies of the students and is not the target of the mentioned learning environments.

Fig. 8. Modeling of a universal plug of minimal volume made of cardboard.

4 An Example of a Problem Solving Approach

Problem solving can be regarded as the endeavor to transform a given initial state into a different, desired target state, regardless of the subject area. This involves overcoming a barrier that exists between the initial and target states [8, 2]. Mathematical problem tasks are particularly characterized by the demanding mathematical concepts or content. The barrier manifests in the inability to apply familiar basic models and solution strategies directly to solve the problem task. This makes it difficult for learners to perform transfer tasks [9]. In the current discourse substantial problem tasks are proposed as a starting point. With these tasks, learners can independently determine the learning resources, the choice of solution methods and representations, the social learning format, and the depth of engagement with the problem field. Substantial problem tasks offer various opportunities for mathematical problem solving, and the individual challenge is primarily controlled by the subject-specific content [9, 11]. In this context, the transfer tasks of learners in mathematical problem solving become particularly evident. The problem of modeling different universal plugs applies on the features of a substantial problem task. It can be used as such a task in mathematical learning environments in classroom and extracurricular mathematical courses.

However, by choosing a problem solving approach in teaching and learning of mathematics, the focus of the mathematical learning environment shifts. It is essentially a form of discovery learning [7, p. 7]. Compared to the direct instruction of relevant concepts and procedures, learners need more time for the elaboration phase. As a result, some of the active mathematical engagement that typically occurs during the practice phase is somewhat brought forward. Important experiences are also gained, and mathematical activities are carried out during the problem-solving phase, which are significant for later understanding [7, p. 7].

To solve problems, it demands problem-solving strategies also known as heuristic strategies, which can be denominated as universally applicable processing rules for a class of problems. One well described heuristic strategy is the variation of representation. This means changing the representation of a problem through a drawing, an action, or a construction [10]. Modeling different universal plugs using different methods is a good example of using the heuristic strategy of variation of representations. It helps in particular dealing with the Problem P by exploring the important finding that there is more than one convex solid matching the conditions of P.

Dealing with Problem P is also a good example for a phenomenon that allow 2D image structure to be interpreted in terms of 3D scene structure. In this way, three points are imported [15]: (1) While the discussion appeals to analytical reasoning, humans readily perceive structure without conscious reasoning. (2) Although we can nicely model several vision cues separately, interpretation of complex scenes surely involves competitive and cooperative processes using multiple cues simultaneously. (3) Our interest is mainly solving a particular application problem in a limited domain, which allows us to work with simpler sets of cues.

5 Conclusion

A well-known geometric problem that dates to the 18th century and was also studied by the Hungarian mathematician Georg Pólya, involves determining a solid with the following properties: The projection of the solid onto the plane is a circle, onto the front is a square, and onto the profile is an isosceles triangle. A handy term is universal plug [14]. Contrary to what the name might suggest there is no unique solution. In fact, there are infinitely many solids that satisfy the mentioned conditions. Assuming, that we limit ourselves to convex solids, these can be ordered by their volume. Thus, a solid with the minimal volume and a solid with the maximal volume can be found, which can be considered as the boundaries of a continuum within which all convex solutions can be identified. Comparing volumes to the volume of a cylinder, which diameter equals its height, it is interesting that the volume of a special non-convex universal plug corresponds exactly to half the volume of the cylinder. Furthermore, the maximal volume is approximately 7.5% larger than half the volume of the cylinder. The approaches to volume determination of these solids can be illustrated through the creation of models using different methods. As a tactile method to illustrate the infinite number of solutions, shaping the models with modeling clay based on a framework with the three shapes (circle, square, triangle) is suggested. A suitable analog method to create the universal plug for younger learners consists in the use of playdough. A framework made of cardboard also helps modelling different versions of the universal plug such as the one of maximal volume. As an analog method creating a non-convex universal plug of volume zero, layering cardboard cuts is recommended. Both methods are suitable for learners of all ages to deal with the given problem.

Acknowledgments. We want to say thank you to PD Dr. habil. Michael Schmitz for giving hints of creating 3D visualizations of the different universal plugs. A special thanks goes to Nicole Polanskij who helped creating analog models and making photographs of it.

References

1. Catel, P.: Mathematisches und physikalisches Kunst-Cabinet, dem Unterrichte und der Belustigung der Jugend gewidme. Berlin (1790)
2. Dexel, T., Witte, A.: Mathematische Problemaufgaben ohne Sprachbarriere. Problemlösen für alle Schüler*innen. DiMawe –Die Materialwerkstatt **3**(1), 55 – 61 (2021)

3. Dörfling, R.: Mathematik für Ingenieure und Techniker. Ein Lehrbuch. Oldenbourg, München (1950)
4. Gradener, M.: The Second Scientific American Book of Mathematical Puzzels and Diversions. New York (1961)
5. Hauser, U., Komm, D., Serafini, G.: Wie Mathematik und Informatik voneinander profitieren können – Teil 1: Abstraktionsfähigkeit. Informatik Spektrum **42**(2), 118–123 (2019)
6. Hemme, H.: Das große Buch der mathematischen Rätsel. Anaconda, Köln (2013)
7. Holzäpfel, L., Leuders, T., Rott, B., Schelldorfer, R.: Schritte zum Problemlösen. PM – Praxis der Mathematik in der Schule **68**, 2–8 (2016)
8. Hussy, W.: Denkpsychologie. Kohlhammer, Stuttgart (1984)
9. Käpnick, F., Benölken, R.: Mathematiklernen in der Grundschule. Springer, Berlin (2020)
10. Komm, D.R., et al.: Problem solving and creativity: Complementing programming education with robotics. In: Proceedings of the 2020 ACM Conference on Innovation and Technology in Computer Science Education (ITiCSE 2020), pp. 259–265 (2020)
11. Krauthausen, G., Scherer, P.: Natürliche Differenzierung im Mathematikunterricht. Konzepte und Praxisbeispiele aus der Grundschule. Klett-Kallmeyer: Seelze (2016)
12. Müller, M., Poljanskij, N.: Gibt es mehr als einen Pólya-Stöpsel? Verschiedene Zugänge zu einer geometrischen Problemstellung. In: Sjuts, J., Vásárhelyi, É. (eds.) Theoretische und empirische Analysen zum geometrischen Denken, pp. 227–242. WTM, Münster (2021)
13. Neubrand, M.: Mathematik zu einem Spielzeug. Didaktik der Mathematik **13**, 63–73 (1985)
14. Pólya, G.: Vom Lösen mathematischer Aufgaben. Einsicht und Entdeckung, Lernen und Lehren. Springer, Basel (1966)
15. Shapiro, L., Stockman, G.: CSE576 Computer Vision. Chapter 12 Perceiving 3D from 2D images. Paul G. Allen School of Computer Science and Engineering, University of Washington, Seattle (2000). https://courses.cs.washington.edu/courses/cse576/book/ch12.pdf
16. Stauss, T.: Frühe Spielwelten. Zur Belehrung und Unterhaltung. Die Spielwarenkataloge von Peter Friedrich Catel (1747–1791) und Georg Hieronimus Bestelmeier (1764–1829). Librum, Hochwald (2015)

The Algorithm Experience at Primary Schools: An Experience Report

Maarten Löffler[1,2]

[1] Utrecht University, Utrecht, The Netherlands
m.loffler@uu.nl
[2] Tulane University, New Orleans, USA
mloeffler@tulane.edu

Abstract. The *Algorithm Experience* is an activity for students of algorithms, in which they take on the role of a computer, and execute an algorithm by hand on a *paper machine*: an unplugged machine that uses envelopes for memory cells and paper cards for values. The experience was initially targeted at high school or university students.

In this experience report, we report on using the activity with primary school children in Louisiana in 2022 and 2023. We discuss several adaptation to the activity that were tried for this purpose and evaluate their success.

1 Introduction

Computational efficiency lies at the heart of the research area of algorithm design. It is of vital importance, since data sets are growing ever larger, and at a much faster rate than advances in hardware; essentially, better hardware cannot keep up with the asymptotic behaviour of a badly designed algorithm. Yet, the same growth in both data size and computation speed is also making the notion of computational efficiency more abstract to end users of those algorithms, and more difficult to explain to inhabitants of a world in which instant computation is the norm.

In 2022, the *algorithm experience* [7] was introduced as a way for students of algorithms, or really anybody, to get a feeling for the notion of computational efficiency by *experiencing* it first-hand. The idea is simple: execute multiple algorithms that achieve the same result, by hand, on the same small input. By doing so, and assuming you make no mistakes, you will notice that you can get to the same results in different ways, and that some take longer than others. In addition, you may feel that some algorithms contain a lot of unnecessary repetition.

The algorithm experience was initially targeted at high school or university students, who take on the role of a computer, and execute an algorithm by hand on a *paper machine*: an unplugged machine that uses envelopes for memory cells and paper cards for values. The experience "unplugged" in the sense that it teaches computational thinking without using (electronic) computers. This idea was first popularized by Bell *et al.* [3]; since then, a large number of such activities covering a wide range of CS topics and aimed at different targets have been developed. See [4] for a recent review.

In the present submission, we report on using the activity with primary school children in Louisiana in 2022 and 2023. In particular, we report on our findings from using the algorithm experience in the context of the BATS and GIST program [1] at Tulane University. In this program, groups of primary school children or primary school teachers visit the university for a day of activities in the STEM fields. We report on adaptations we made to the experience to make it suitable for primary school children, and the evolution of the activity through four iterations of BATS and GIST.

We describe the original activity in more detail in Sect. 2, and then discuss the adaptations that were made in Sect. 3. Finally, we reflect on our experience in Sect. 4.

2 The Algorithm Experience

For context, we begin by briefly reviewing the algorithm experience as it was developed before 2022. The algorithm experience consists of executing multiple algorithms for the same computational problem on the same input, by hand. These algorithms, and instructions on how to execute them, are described in instruction booklets; see Fig. 1 for an example.

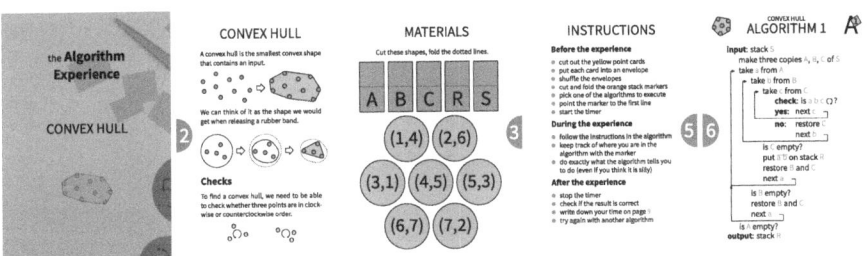

Fig. 1. Some pages from the instruction booklet for the *convex hull* problem.

2.1 Problems and Algorithms

The original version of the experience targets the *convex hull problem*, arguably the most central problem in the field of geometric algorithms. It was selected because of its central role and visual appeal - to a human, it is immediately clear whether the output is correct or not. In the convex hull problem, we are given a set of points in the 2-dimensional Euclidean plane, and need to compute the smallest convex set that contains them. The activity compares two algorithms: a simple brute-force algorithm that runs in $O(n^3)$ time, and the algorithm of Andrew [2] (an adaptation of Graham's scan) with some minor adaptations out of practical considerations, which runs in $O(n \log n)$ time; refer to [8] for more details of these algorithms and an early manual execution report.

Later iterations simplified the problem to computing just a single edge of the convex hull (namely, the topmost edge that crosses a given vertical line); this is computationally easier and already sufficient to prove the point. The advantage is that the algorithms are simpler to execute and take less time; however, a disadvantage is that the problem itself is a bit less natural and harder to explain, making the activity somewhat more difficult to use.

The algorithms for both problems were originally provided in pseudocode. In later iterations, the experience also experimented with providing the algorithm in graph (or flow chart) form to make it more accessible for participants not familiar with coding, although the graph representation still resembled the layout of the original pseudocode. Refer to Fig. 2 for an example of two different layouts of the same algorithm for finding the topmost edge of the convex hull that connects a blue point (the points left of the y-axis were "blue") to a red point (the points right of the y-axis were "red").

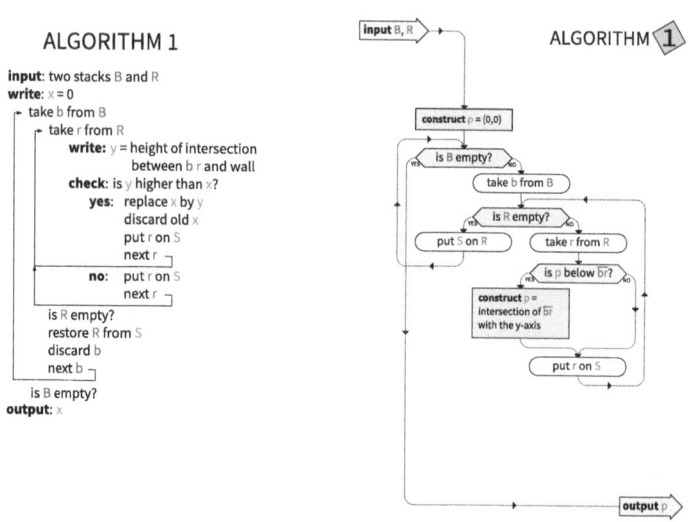

Fig. 2. The same algorithm to find the topmost edge of the convex hull, formatted as pseudocode (left) or as a graph (right). (Color figure online)

2.2 The Paper Machine

Participants execute the algorithms on the *paper machine*: a simplified model of a computer that simulates the concepts of memory and values, as well as stacks (arrays) and pointers.

Specifically, the machine consists of an (in principle unlimited) number of *memory cells*, each of which can store a single *value*. These cells are in turn stored on (a limited number of) named *stacks*. Finally, the machine supports a small number of named *pointers* that point to specific memory cells. When a cell has a pointer to it, it does not need to be on a stack, and only when a cell has a pointer to it, the value stored in the cell can be inspected.

Physically, memory cells are represented as *envelopes*, and vales as *paper cards* that fit inside the envelopes. Stacks are static locations that can support a stack of envelopes, and pointers are small paper pointers that can be placed on (opened) evelopes. Refer to Fig. 3.

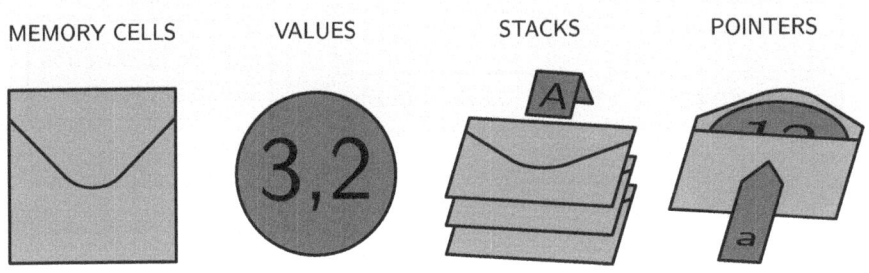

Fig. 3. The elements of the paper machine.

This physical model simulates the concept of memory cells of a computer, and the idea that a computer does not "know" or "remember" which numbers it has seen before: when an envelope is closed again, its value is "forgotten" and to "remember" it, we need to obtain a pointer to the envelope again.

The values stored in envelopes can in principle be of different types, but the concept of types is not very prominent in this activity; in fact, in the original activity all values were of the type "point in 2-dimensional space", although the in later iterations the type "real number" was also used to temporarily store information. Dealing with multiple types in the same algorithm arguably makes the activity more complex to execute, though.

Formally, our machine supports *manipulation*, *predicates*, and *synthesis* operations. Refer to Fig. 4.

Manipulation Operations. Most instructions in an algorithm will be *manipulations* of the state, which rearrange envelopes without inspecting or altering their values at all. There are, for example, moving stacks of envelopes to other stacks, taking a single envelope from a stack (and obtaining a pointer to it), or putting an envelope on a stack (and losing the pointer to it).

Predicates. An algorithm also needs *predicates* (or *tests*). Predicates can either be on the memory level (such as testing whether a stack is empty or not), but also on the value level: whenever several pointers are pointing to different envelopes, it is possible to perform predicates on the *values* stored in those envelopes.

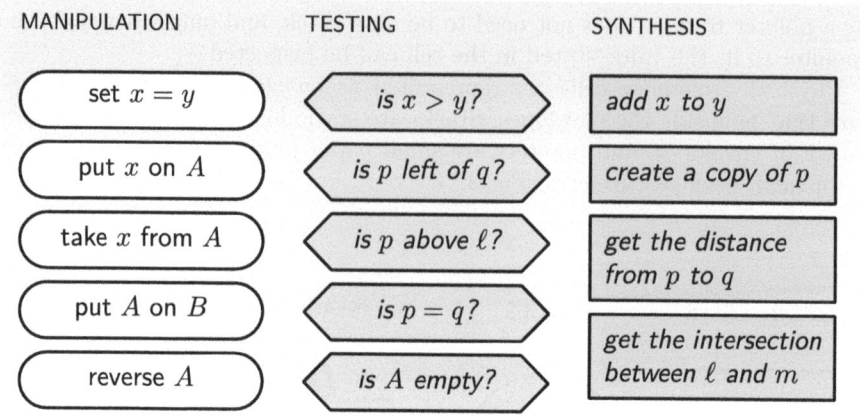

Fig. 4. Examples of operations on the paper machine.

Synthesis. Finally, *synthesis* operations (or *constructions*) allow the creating of new values. These are arguably more complex to perform than manipulations or predicates. Furthermore, they are destructive in the sense that they require physical writing of new values on paper cards. Therefore, the original activity avoids using synthesis operations as much as possible.

3 The Algorithm Experience for Primary School Children

In the fall of 2022 and spring of 2023, we had the opportunity to use the algorithms experience in the context of the BATS and GIST program, in which groups of primary school children (or groups of teachers of primary school children) visit Tulane University to participate in several workshops. The program specifically targets children from underrepresented demographic groups and children who otherwise have reduced contact with university education. For this purpose, the activity needed to be self-contained and possible to execute in 50 min.

When adapting the activity for primary school children, we need to ensure that it is simple enough to be easily explained to children without any prior experience with programming or the idea of algorithms, while still achieving the main goal of experiencing different algorithms. Specifically, the issues mentioned above such as avoiding mixed types or avoiding synthesis operations become even more important.

In total, the activity was done 12 times, with 10 groups of children and 2 groups of teachers; the group sizes varied from 6 to 12 participants. In this section, we review several aspects of the experience that we altered.

3.1 Problem and Algorithms

Although the convex-hull and single-edge-of-the-convex-hull problems are in principle nice problems because of their visual appeal, we decided against using them in this case. There are two reasons for this. First, the problems are still somewhat complicated to explain, and it needs to be completely clear for the activity to make sense. Perhaps in the context of multiple sessions, it would be possible to cover these problems, but in a self-contained activity we felt it would take up too much of the time. The second reason is that the algorithms do use some synthesis operations or multiple types, which we would like to avoid as much as possible.

Instead, we decided to base the activity on arguable to most fundamental of all algorithmic problems: sorting. This feels like a good problem both because it is immediately clear what the objective is, and because for children for whom this is their first encounter with the concept of algorithms.

For the problems of sorting, we also chose two algorithms to execute, one that is not efficient and one that is. For the non-efficient algorithm we picked *selection sort* [6], because it is simple and only relies on basic predicates, and can be represented as a simple pair of nested loops. For the second algorithm, we picked *merge sort* [5], mainly because it can be also expressed in a non-recursive way and still be relatively simple. (Recursion is difficult and not very intuitive to emulate on the paper machine.)

One disadvantage of our choice is that non-recursive merge sort does require one to maintain a counter that keeps track of the level at which the algorithm is currently operating. We opted to solve this by keeping an integer variable x that is not stored in an envelope but just remembered, which starts at value 1 and is doubled a few times. This is simple enough that it works, but still led to significant questions, so it would be better to avoid it if possible.

3.2 Presentation of Algorithms

For younger children, using pseudocode is not really an option, so we expanded on the graph view in Fig. 2. In order to make it easier to keep track of the current state of algorithm, we represent each node of the graph as a large circle. At the start of the activity, a wooden pawn is placed on the "input" node, and this pawn can be moved through the graph. We also found that the larger nodes make the graph more appealing to traverse, which is increased even further by arranging the nodes in a more playful way.

In order to make the graph as easy as possible to traverse, we tried to phrase the instructions inside each node as intuitive natural language. That is, we do not attempt to formally introduce the paper machine, but rather describe operations only where they are needed. We also tried to group instructions into a single node when possible to reduce the size of the graph, such as "take all envelopes from T, flip the stack, and put them on S". However, this instruction turned out to be too complex, so was later split again into two separate instructions: "flip T (pick up the whole stack and put it back upside down)", and "put T on S (pick up the whole stack and put it, in the same order, on S)". Throughout the different iterations of the activity the phrasing as been adapted based on earlier experience; striking a good balance between the size of the graph and the clarity of the instructions is an important aspect of the design. Figure 5 shows two iterations of the same algorithms (selection sort on the left and merge sort on the right) and their layout.

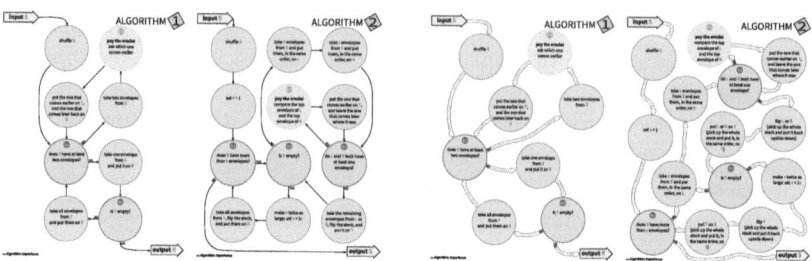

Fig. 5. Layout of the algorithms as used in BATS (left) and GIST (right).

3.3 Measuring Runtime

Another challenge is how to measure the performance of different algorithms. Measuring actual time is not ideal, since the second algorithm is more complex to understand, so participants take more time per instruction than in the first algorithm. Furthermore, the first algorithm is more repetitive, so as time progresses participants tend to become faster. On the other hand, counting the total number of instructions performed is tedious and would slow down the activity quite a bit.

We opted instead to only measure comparisons. Counting comparisons is actually well understood in the analysis of algorithms, and while this is not quite the same as measuring total instructions or runtime [9], it does seem to be a reasonable metric to get the point of the activity across.

For the size of the input, we opted for a set of 8 items. In this case, the number of comparisons needed by selection sort is $8 \cdot 7/2 = 28$, while merge sort requires between 13 and 17 comparisons depending on the initial permutation. This difference is big enough to make the point, while the numbers are small enough to make it easy to keep track of (and keep the duration of the activity low).

3.4 Introducing the Oracle

The original algorithm experience can be monotonous to do, and requires continuous good concentration to avoid making mistakes. While this is part of the objective (to show that some algorithms are repetitive), it does make it less suitable for children.

To alleviate this, we decided to let participants work in pairs, with one participant taking on the role of the *computer* and the other the role of the *oracle*. Both participants are sitting side by side, but each have distinct roles and responsibilities. Refer to Fig. 6.

The Algorithm Experience with Primary School Children   111

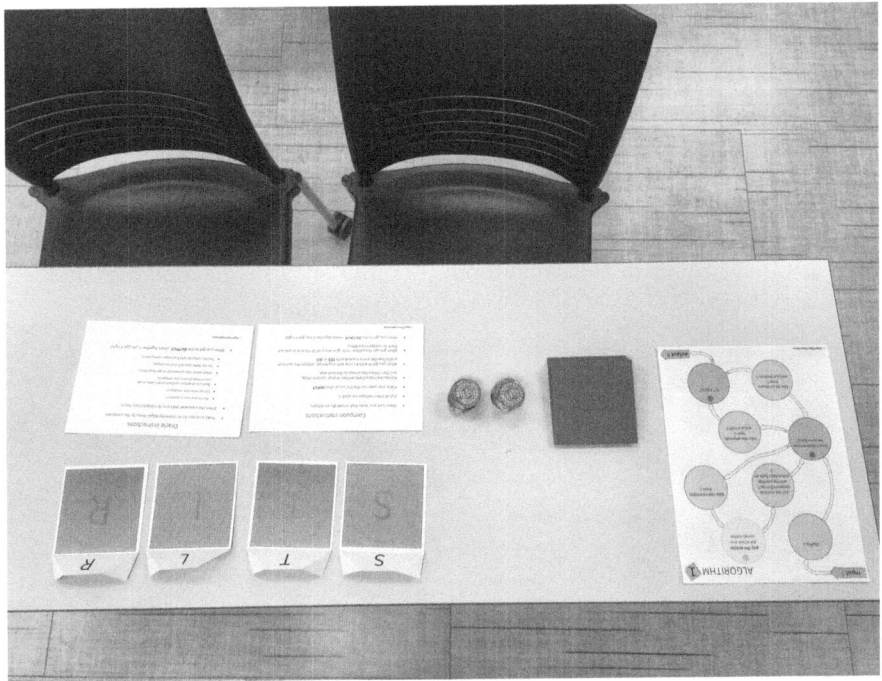

Fig. 6. Setup for two participants.

The computer executes the algorithm, while the oracle is in charge of the actual comparisons. To count comparisons, the computer has to explicitly pay a gold coin to the oracle for each comparison. The computer is never allowed to look inside the envelopes, and the oracle is never allowed to move the envelopes or the pawn.

Working in pairs makes the activity much more fun and kept participants more engaged, and additionally provides an opportunity for participants to check each other, making mistakes less likely to occur. Furthermore, we observed some participants playing into the "paying money" aspect and felt this as a real measure of how expensive an algorithm is to execute.

Finally, working in pairs allows for switching roles between the execution of the first and the second algorithm, which provides a welcome variety.

3.5 Values

One aspect about the activity is which values need to be sorted. When sorting, for instance, numbers, we have a clear idea of what the correct order is, but this also makes it tempting to "cheat": participants may remember which numbers are in play and what their order is, the computer may not want to pay the oracle for information they feel they already have. This is even more clearly the case when the numbers being sorted are the numbers 1 through 8: then even the required position is already clear from the number. On the other hand, sorting a random set of numbers may feel less clean.

To avoid these issues, we let our values be items that are not inherently ordered (namely, vegetables). The oracle has access to a secret sheet of paper that contains the items in the correct order, to be able to compare items when requested, but otherwise neither participant has the means to compare items.

Additionally, it has the benefit that the order (and indeed the set of items itself) can be different for each pair of participants, making it no issue if they overhear other groups in the same room. Figure 7 shows three different lists.

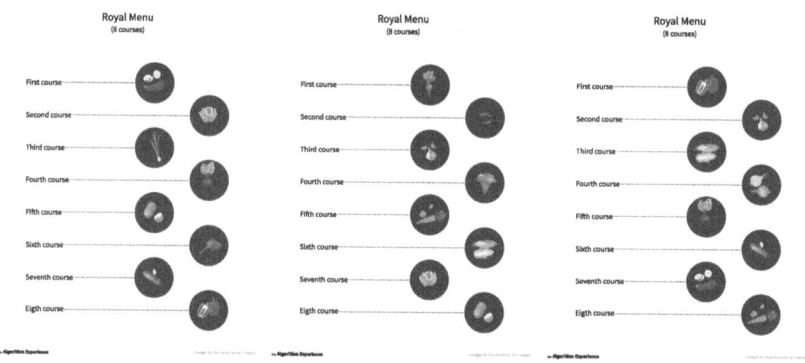

Fig. 7. Randomly generated orders.

3.6 Dealing with Mistakes

Although working in pairs reduced the chance that participants make mistakes, we still observed many mistakes being made during the execution of the algorithms. The reasons for these varied from misunderstanding the activity or misunderstanding specific instructions, to mistakes in performing predicates or forgetting to move the pawn forward on the board.

Making mistakes during the execution of the algorithms can potentially have quite a large effect on the effectiveness of the activity: if the algorithm does not even produce the correct output, it is hard to appreciate its efficiency. Luckily, sorting items does give the possibility to checking if the output is partially correct: if most items are in the correct order, then a human participant may still feel they did a good job, even if the link to actual theory of computing is not technically there anymore. In this respect, selection sort is quite robust: even if one incorrect item has been selected, the algorithm will still correctly sort the remainder. In contrast, merge sort is not as robust: one can be unlucky when merging two almost-sorted subsequences that are not being merged properly.

The separate sheet with the "solution" also makes for an exciting moment at the end of the activity, when participants check their output one by one.

We noticed great differences in the number of mistakes that were made between individual participants, as well as on average between different schools. To reduce the number of mistakes, it helps to make sure the instructions are clear, but also to make participants aware that it is ok to make mistakes and to try to repair them, when they notice, or even to start over if it happens during the first few minutes - generally, the longer the activity continues, the less likely mistakes are (unless participants get bored or tired, which can increase mistakes). Finally, it helps to have enough supervisors in the room that can notice mistakes being made and act if necessary.

4 Discussion

All in all, we have been positively surprised by the effectiveness of the Algorithm Experience on primary school children. Although they are more challenging to work with that high school or university students and need more guidance, most of the participants did manage quite well and did appear to take home the idea that computers can do very tedious work, and that being smart about it helps. We feel that the adaptations that were made do make the activity suitable for younger children, although more improvements can still be made.

Perhaps as a main conclusion, we can say that primary school children are not too young to learn about algorithm analysis, and given the huge impact that algorithms will have on their lives, they probably should be learning about this - even if just for 50 min.

4.1 Future Work

Based on our experiences, we plan to refine the activity more for use in primary schools. Most importantly, it would be good to select two algorithms that are both purely predicate-based with a uniform value type; for this target audience we feel that is more important than being true to asymptotic optimality. Another point of interest is the clarity of the layout of the algorithms, and the robustness to mistakes.

Aside from the pragmatic solutions that we sought, and that will be necessary to refine the activity further, we also believe that our experiences can guide theoretical algorithms research. What algorithms can be represented most succinctly as a instruction graph that can be executed on the paper machine? What algorithms are both efficient and robust to mistakes in predicates, or mistakes in following steps of the algorithms?

Acknowledgements. We are grateful to Irina Kostitsyna, Marjolein Haagsman, Michelle Sanchez, all previous participants in the experience, and anonymous reviewers for their very valuable and highly appreciated feedback.

References

1. BATS and GIST. https://sse.tulane.edu/teacher-gistbats-pd
2. Andrew, A.: Another efficient algorithm for convex hulls in two dimensions. Inf. Process. Lett. **9**(5), 216–219 (1979)
3. Bell, T., Witten, I.H., Fellows, M.R.: Computer Science Unplugged - an enrichment and extension programme for primary-aged children, 2002. csunplugged.org
4. Huang, W., Looi, C.: A critical review of literature on "unplugged" pedagogies in K-12 computer science and computational thinking education. Comput. Sci. Educ. **31**(1), 83–111 (2021)
5. Katajainen, J., Träff, J.L.: A meticulous analysis of mergesort programs. In: Bongiovanni, G., Bovet, D.P., Di Battista, G. (eds.) CIAC 1997. LNCS, vol. 1203, pp. 217–228. Springer, Heidelberg (1997). https://doi.org/10.1007/3-540-62592-5_74
6. Knuth, D.: The Art of Computer Programming, Volume 3: Sorting and Searching, pp. 138–141, Third edn. Addison-Wesley, Boston (1997)
7. Löffler, M.: The algorithm experience. Presented at the 6th International Conference on Creative Mathematical Sciences Communication, 2022
8. Löffler, M.: A manual comparison of convex hull algorithms. In: Proceedings of the 34th Symposium on Computational Geometry (2019)
9. Peczarski, M.: New results in minimum-comparison sorting. Algorithmica **40**(2), 133–145 (2004)

Curricular Desicion-Making

Curricular Choices for Computational Thinking in Large Scale Low Resource Environments

R. Ramanujam[1] and Vipul Shah[2](✉)

[1] Azim Premji University, Bengaluru, India
jam@imsc.res.in
[2] Tata Consultancy Services, Pune, India
v.shah@tcs.com

Abstract. The Indian National Education Policy 2020 advocates the introduction of Computational Thinking (CT) and coding in school, and the National Curriculum Framework 2023 follows through on this. Undoubtedly, given the scale of this effort (reaching a quarter billion children), the challenges are aplenty.

As it happens, India already has some *CS Unplugged* style experience with large-scale initiatives in introducing CT in low-resource environments, necessarily working with the educational machinery of state governments. We discuss these experiences, focusing on the three dimensions of *large-scale*, *low-resource*, and *state-run* nature of CT education programmes. We identify some core challenges in the introduction of CT in schools that are specific to these contexts and articulate some questions on curricular choices.

Keywords: Computer Science · Computational thinking · K-12 · Education

1 Background

In India, computer science as a discipline is taught principally in universities, with some preparation for it at the higher secondary level. In the first ten years of school, so-called *computer classes* have tended to be mainly on the usage of computers, platforms, and the Internet. Even this is mostly in urban private schools. All this is set to change, with the advent of the National Education Policy 2020 [9], which advocates Computational Thinking (CT) and coding throughout the school years. The National Curriculum Framework 2023 [14] lists the development of CT among the curricular goals at the middle, secondary, and higher secondary stages. With all this push, it is clear that India will soon see systematic incorporation of CT and coding in the school curriculum. Undoubtedly, given the scale of this effort, reaching a quarter billion children, the challenges are aplenty.

Principal among these challenges for computing education is the so-called digital divide. According to the Oxfam India Inequality Report, 2022 [16], only about 9% of students enrolled in *any* course of study had access to a computer with internet and only 25% of enrolled students had access to the internet with any kind of devices. This is especially stark when it comes to rural vs urban usage. Only 31% of the rural population of India uses the Internet whereas 67% of the urban population uses the Internet. Further, according to the report, socioeconomic factors such as gendered social norms, affordability, geographical location and levels of digital literacy determine who owns and gets to use the available gadgets.

As it happens, India has already some experience of large-scale initiatives in CT that ACM India has been a part of. Given this localised context, and the obvious challenges ahead, we articulate some questions that may be of interest to the larger computing education community and point to some tentative answers. What we present is a practitioner's perspective, of a curriculum based on positions articulated by state policy documents. The obvious limitation of this perspective is its lack of foundation on empirical research study. It is hoped that the account of implementation in low resource contexts would nevertheless be of interest to the community.

2 The Localised Context

ACM India's formal initiatives on computing education began in 2016 with the establishment of **CSpathshala**, an effort to bring computational thinking to schools [2], in the style of *CS Unplugged* [4].

2.1 CSpathshala

CSpathshala was launched with the objectives of creating awareness of CS education, building a draft CT curriculum, training teachers, creating a community of practice and influencing policy on CS education. An overview of the CSpathshala curriculum is provided in Sect. 3.3. To date, over 20,000 teachers have been oriented on CT through training programs and over 300,000 students have been learning some aspect of CT every year, largely in rural government-run schools.

To create and support a community of practice, CSpathshala has been organizing CTiS [1], an annual conference for teachers. It provides a platform for teachers to share experiences and best practices in CT classrooms. The conference also serves as a window into CT implementation in the country. Experiences shared by teachers at the conferences indicate that students have received the program enthusiastically, though there has been no formal impact study as yet.

2.2 Andhra Pradesh

The Department of Social and Tribal Welfare in the state of Andhra Pradesh runs 425 residential and day schools (classes 5–12) for students from socially and economically oppressed sections of society. In 2017, the department started an initiative called *Naipunya Vikasam* to train students on communication skills, computational thinking skills and digital fluency. The majority of the students are first-generation school-goers belonging to marginalized communities with an annual family income of less than EUR 1,100. This effort reached nearly 200,000 students from the poorest sections of society each year. The implementation helped us better understand the needs of students from rural areas, especially in terms of challenges faced by teachers and the need for greater localization and contextualization.

[17,18] share experiences of teaching CT in these welfare schools. Feedback from both parents and students has been extremely positive, but the government has not taken up any formal impact study.

2.3 Tamil Nadu

Tamil Nadu state initiated a process of curricular reform in 2017, seeking to formulate a forward-looking school curriculum for 21st-century needs. The position paper on mathematics education [21] asserts that "... a strong foundation for computational thinking will be essential for children growing up in this century". Further, it claims that "such understanding and thinking lies squarely within the realm of mathematics in school". Subsequently, in the new syllabus [22] formulated in 2018, elements of CT were explicitly incorporated into mathematics education. For classes 1 to 5, this came in two new tracks called Patterns and Information Processing (IP) and for classes 6 to 8, in tracks called Modelling and Information Processing. The latter included Systematic Counting, Data organization, Iteration and Devising and following algorithms [Sect. 3.3].

Textbooks based on the new syllabus were planned for a three-phase implementation, to get teachers trained and acclimatized to the new elements. Though it was rolled out in 2019, due to the pandemic the syllabus was fully implemented only during the 2022 academic year. However, considering that this reaches nearly 30,000 schools comprising close to 2 million children taught by nearly 80,000 teachers, the system constitutes one of the largest CT education experiments at the elementary stage.

The central characteristics of this system are: decoupling coding and digital literacy components from CT and integrating the rest into mathematics education, using data as the principal vehicle (so that it may be related to the prevalent understanding of data handling among teachers). The use of mathematics teachers, well versed in pedagogy, was seen as preferable to using existing computing teachers who were educated only in computer science and not on pedagogy.

While there has been no publication from the state government on any impact study, the feedback from teachers has been highly enthusiastic and positive [10].

3 Curricular Implementation in Low-Resource Environments

Typically curriculum design takes place within a policy framework that is expected to guide the implementation of curriculum. Curriculum implementation refers to how teachers deliver instruction and assessment through the use of specified resources provided in a curriculum. Curriculum designs generally provide instructional suggestions, scripts, lesson plans, and assessment options related to a set of objectives.

3.1 Curriculum Implementation Frameworks

According to Fullan [6], *implementation consists of (1) using new materials, (2) engaging in new behaviours and practices and (3) incorporating new beliefs.* In the context of CT, each of these components acquires significance. Bringing CT into the national school curriculum involves action along all these dimensions. While new curricular materials exist, engaging in new behaviours and incorporating new beliefs will call for a great systemic shift.

In all this, the role of **availability of educational resources** requires special attention. In low-resource classrooms, teacher beliefs and expectations are limited by their perceptions of what can be achieved in the classroom, even when they are tuned to student aspirations. Rittel and Webber [20] refer to this as a wicked problem since teachers must teach the students different subjects without the proper resources in their classrooms to do so. According to them, social policy on such problems is *bound to fail*.

In the context of Computational Thinking curricula, there is often an implicit premise relating to the availability of computing devices and appropriate software platforms on which children can learn programming ([5]). Another implicit premise of CT curricula is the availability of teachers educated not only in computing but also in the core aspects of school education and the psychology of children's learning. ([23]).

There is one natural answer to these premises. If devices and teachers are needed, simply ensure that they are available before you start implementing the CT curriculum. The Global North has largely followed this answer. In low-resource contexts, this is hardly a reasonable premise. Curriculum design in India can ignore neither the small section of children who have access to devices nor the majority who do not.

Is it then only a matter of *bridging the 'digital divide'*? Structural inequity in the system is not overcome so easily. It is not only a matter of making the devices available but also aligning pedagogy with the use of devices. Moreover, teachers have to "buy-in" to the changed pedagogy, and internalise its use in the classroom [15]. When it comes to connectivity, there are further problems relating to restricted and safe use in the classroom. After all this, ensuring that this change achieves the intended curricular objectives is difficult as well. It is fair to say that in a country like India, there is little systemic readiness for such reform.

On the other hand, we can revisit the curricular aims of CT in schools in the light of these constraints, and in the process, find ways of overcoming the challenges, at least at the preparatory and middle stages of school education.

3.2 Curricular Aims

The surface objectives in CT education relate to the use of computing platforms, tools and devices. This implies not only mastery over device use, but also creating in students a predisposition to identify and utilise educational contexts in which the use of computation can help, and employing computations accordingly. In this sense, CT is a mode of thinking that helps the student to articulate problems and solutions in a way that makes them amenable to solving by computer. An example of this would be plotting the trajectory of a ball under Newton's laws.

The deeper objectives relate to nurturing the inner resources of the student and thus address the principal goals of education *per se*. From this viewpoint, computation assumes focus (rather than computing devices) and developing thinking that *underlies* computation becomes the educational goal [14]. Such thinking helps the student understand data organisation, scaling, assessment and comparison of procedures, iteration and modular abstraction, independent of whether solutions are sought by computer or implemented on computing platforms. An example of this would be knowledge of multiple procedures for integer multiplication or division, and an understanding of which is best used when.

Developing critical thinking, autonomous learning and resource consciousness are central aims of modern school education, as also preparing the young for participation in a modern democracy. Viewed thus, the aims of CT education at school are about utilising the tremendous new potential brought by computation for autonomy and empowerment, and at the same time developing a critical outlook on data and algorithms, and sensitivity to resource use as practices scale up [7]. These guiding principles are in tune with the 'inspirited curriculum' orientation of [3].

3.3 Curricular Components

According to [8], the key concepts of CT are: logic and logical thinking, algorithms and algorithmic thinking, patterns and pattern recognition, abstraction and generalisation, evaluation and automation. They list CT practices as: problem decomposition, creating computational artefacts, testing and debugging, iterative refinement (incremental development) and collaboration & creativity. These concepts and practices were considered from the dual prism of implementation in contexts lacking computers and where CT would be introduced predominantly through integration into mathematics curricula. The considerations were concretized into the following components. Specifically, we highlight below the potential for creative pedagogy, in the spirit of the CMSC conference.

1. **Scaling**: Systematic listing and counting of relevant parameters and verifying that all have been counted is essential for the transition from additive reasoning to multiplicative reasoning. This also paves the way for functional variation and *the use of symmetries for counting*. Scaling prepares students for understanding how difficulty of solutions could grow with problem size.

2. **Iteration**: Looking for patterns, finding a mechanism for pattern generation and modification, and visualisation of new patterns are all not only enjoyable experiences but also provide *a link between aesthetics and formal reasoning*. Understanding the power of iterating simple rules creates a foundation, not only for computation but also for the study of dynamics of systems.
3. **Data organisation**: Data can be represented in multiple ways, and which one is to be chosen when depends on the use such data is to be put to. Moreover, storage and retrieval of data requires memory structures. Designing such data organisation is neatly coupled with an understanding of scaling and iterative data access.
4. **Modelling**: Discrete modelling of problems from real-life situations is largely unfamiliar territory in Indian schools. Discrete structures like lists, trees, maps, graphs, lattices and networks arise naturally and provide abstract problem spaces for computation. Working with concrete representations of such structures early on can help in creating mental models for computational abstractions. This again has links with *aesthetics*.
5. **Algorithms**: Starting from two-digit addition in arithmetic, school education provides a variety of algorithms for students to learn – so much so that mathematics or science education often degenerates into a mere memorisation of pre-set procedures to be enacted on specific numerical data. *Following* algorithms with a view to understanding them is no doubt desirable, but *devising* procedures is at the heart of computational thinking. This requires a facility with procedures, reasoning about them, consideration of procedural alternatives and selection among them based on a clear rationale.
6. **Programming**: Concrete implementation of data organisation and algorithms on specified platforms to solve given problems is an essential componet of CT. Coding, when accompanied by a feel for program structure, can be exhilarating while coding as the translation of informal computing into a given formal language can be painful. Hence creating a good disposition for programming early on is essential for achieving eventual fluency.

3.4 Stagewise Curriculum

According to the NEP 2020, learning stages have been identified as: *foundational*: comprising the kindergarten and the first three years of school, *preparatory*: comprising the next two years, *middle*: grades 6 to 8 providing a transition to secondary school, and *secondary*: the last four years of schooling. An overview of the curriculum at the various developmental stages is show in Fig. 1.

Foundational

This is a stage of celebration, where the joy of living and childhood predominate. Opportunities for learning, according to the NEP, consist of "alphabets, languages, numbers, counting, colours, shapes, indoor and outdoor play, puzzles and logical thinking, problem-solving, drawing, painting and other visual art, craft, drama and puppetry, music and movement." These are all opportunities

	Systematic Counting and Rigour	Patterns	Algorithms & Programming
Foundational (grades 1-3): Embodied CT - with self and social spaces	Counting methodically: How many items do you have? How do you know you have counted them all? What was the last numbr you counted? Categorising objects based on visual attributes such as shapes and colours	Patterns in daily life, visual patterns, patterns in numbers, music	Carry out (follow) instructions for getting from one location to another using small instruction set. Gradually introduce conditionals & loops
Preparatory (grades 4-5): Concrete material, concept or content driven	Puzzles that can be solved through systematic analysis. Count number of combinations, with and without constraints. Read, construct, and analyze tables, maps and other data structures.	Spot similarities among shapes and numbers, word and number sequences, symmetry	Devise precise instructions, discover & express basic sorting algorithms, conjunctions or disjunctions of basic instructions.
Middle School (grades 6-8): Explore abstractions, new world views	Counting combinations – moving from enumeration to other methods. Complex puzzles that can be solved through systematic analysis. Encrypting and decrypting messages. Binary encoding	Simple mathematical activities using patterns, Using repetitive process in tasks like sequential search	Searching & Sorting, use flow charts to describe algorithms; simulate algorithms, optimizations

Fig. 1. CT Curriculum

for **embodied computational thinking and practice**. Working with concrete manipulatives, children's own bodies and social spaces is necessary ([11]).

Systematic Counting and Rigour: While counting has always been part of school education, the emphasis has been on children learning to count in whichever way. This is necessary but it also provides an opportunity to discover systematic counting and enumeration, which is crucial for CT. Categorising objects based on visual attributes such as shapes and colours also helps. While teaching these topics stress is laid on rigour, by asking children to verify whether they have missed out any cases, and how they know if all possible cases have been accounted for. As an instance, the activity of asking a child to distribute 20 toffees among 5 children is followed up by asking whether all children got an equal amount, and how they know this. Such reasoning about procedure lies at the heart of CT.

Patterns: The ability to spot patterns and express their characteristics is an important skill that a child needs in order to develop computational thinking. Children in this age group are taught to spot patterns in objects that they encounter in their lives. *Rangoli* or *kolams* (culturally rooted practices that use abstract patterns of dots and curves in drawings made on the ground with rice flour), patterns in leaves, patterns in dress, upholstery etc. are some examples of visual patterns that a child is likely to encounter daily. Such culturally responsive practices bring CT closer to children's lived reality [12]. Children are also encouraged to identify and express patterns that they have not been exposed to earlier.

Algorithms and Programming: Children at the foundational stage are introduced to programming by providing them with a limited set of instructions to work with. As an instance, an *unplugged* exercise uses instructions written on cards: one child enacts a robot that follows instructions given by another child. The instruction set is very small and includes only directions such as move forward, turn right and turn left. Children are asked to provide a sequence of such instructions that will take the robot from one place to another while avoiding obstacles. This activity is used widely and has been shown to be effective in teaching programming to younger children. As we have argued at length earlier, such 'unplugged' activities constituting embodied learning are necessary in low-resource contexts. The number of instructions is gradually increased to include conditionals and loops. Children are asked to identify repetitive sequences of instructions, which prepares them to identify functions at a later stage of learning.

Preparatory
At this stage, while the principal mode of learning continues to be activity-based, experiential and using concrete material, concept or content driven activity is also initiated ([14]). Thus, systematic listing, counting and reasoning about counts, iterative patterns, multiple data representations, devising and following algorithms and basic program structures to create artefacts are explored.

Systematic Counting and Rigour: At this stage we introduce children to puzzles that can be solved through systematic analysis. Some examples are 4×4 Sudoku, Arithmagons, Towers of Hanoi etc. They are also taught to count the number of combinations, the number of squares in a grid, words in a grid etc.

Patterns: Children in classes 4 and 5 are taught to spot similarities among shapes and numbers. They are also able to spot patterns in word and number sequences, and are introduced to more complex concepts like symmetry.

Algorithms and Programming: Children are taught to discover and express algorithms for towers of Hanoi, for finding minimum from numbered cards, and for physically sorting cards based on number. They can handle more complex instructions to a 'human robot'; these could be conjunctions or disjunctions of basic instructions.

Middle
This is a stage of consolidation, building on foundational literacy and numeracy, to explore abstractions and structure across curricular subjects, and to discover connections and new ways of looking at the world ([14]). Indeed, this stage provides the foundation for the major potential of CT education outlined earlier, and all the curricular components of CT are relevant across the subject areas.

Secondary
CT emerges as a discipline at this stage, with a view towards both providing essential skills for 21st-century living for those who exit the system after compulsory school, as well as providing a computational foundation for higher

education. The social goals of CT education, multi-disciplinarity of CT in providing tools for exploration across disciplines, and the power of programming are emphasised at this stage. All curricular components are relevant, with some components accentuated based on students' choice of subject disciplines. Student autonomy and attention to life aspirations are essential at this stage.

We do not have implementation experience of CT at the secondary stage of school.

3.5 Curricular Outcomes

The formulation of explicit *milestones* in terms of curricular outcomes at every stage cannot be done without consideration of the status of schools, availability of educational resources and adequately qualified and prepared teachers. However, certain broad indicators can be provided as follows, in line with [14].

At the end of the foundational stage, every child should be *exposed* sufficiently to gain familiarity with core components of CT such as recognition of patterns and symmetries, multiple ways of counting and count verification, enactment of algorithms and repetitive processes, and physical arrangements of embodied data.

At the end of the primary stage, every child should have *experiential fluency* of the components listed above, continuing from the foundational stage. The child should be able to assess procedures in context and compare different procedures, again in context. All CT components (except with regard to devices) must be familiar to the child in a *contextualised* setting,

At the end of the middle stage, every child should have a basic foundation and a rudimentary formal understanding of all CT components. Familiarity with basic program structure and coding is also required.

At the secondary stage, the emphasis is on mastery of selected CT components, depending on the student's interests, aspirations and choice of subject grouping. Milestones are articulated accordingly.

3.6 Pedagogy

The NEP 2020 articulates the goal of pedagogy to be "to move the education system towards real understanding and towards learning how to learn – and away from the culture of rote learning as is largely present today." This is entirely appropriate for CT pedagogy as well, and needs to be the core guiding principle.

In the context of CT, processes such as pattern recognition, visualisation, abstraction, representation, heuristic strategisation, trial and error, guess and verify, argumentation and formal communication are all important and contribute essentially to the development of CT in the learner. The use of board games, audio-visual games and kinesthetic games, singly and in groups, provide excellent opportunities for foregrounding these processes.

Processes specific to CT relate to discrete modelling, programming and data organisation. While constructivist pedagogy can surely apply to these contexts as well, research is needed on the design of appropriate pedagogic tasks that provide a low threshold and high ceiling for learners engaging in the task from multiple layers of background and capability.

Finally, pedagogy needs to be guided by the principle of **inclusion**. Engaging every child with a sense of success has to be a matter of right and principle. CT can effectively address the needs of *children with disabilities*, with the use of computing technology tools providing special educational opportunities. Indeed, learning from such practices can aid pedagogy for all classrooms.

4 The Principal Standpoint, and Questions Arising from It

A central question in computing education at the school level relates to the use of devices and teaching programming. Across the world, many countries are going for computing education at all levels. In the Indian context, as in the case of many Asian and African countries with a large proportion of low-resource classrooms, many specific issues assume the centrality of concern.

- Devices: Only a small fraction of students have access to a computing device, individually or shared. With the exception of a small elite section, the situation is largely similar in private schools as well. However, this elite not only has individual smartphone access with connectivity but often tablets as well, enabling the use of many technological tools for learning. In this scenario, there is a grave danger that the push for universalisation of CT and computing education may only further deepen the digital divide.
- Curricular space: Should computing be taught as a specific subject in all years of school? Computing is viewed as a separate subject in school, with its own curriculum, time periods, assigned teachers, and assessment modes. The alternative of integrating aspects of CT into the existing structure unless/until educational imperatives dictate separation, requires study.
- Teachers: With teacher education programmes being largely disjoint from computing education in the universities, the current pool of persons trained in computing, as well as school-level pedagogy, is miniscule. While teacher education curricula spell out pedagogic practices in great detail for the existing school system [13], this is yet to be done for CT.
- Assessment framework: Existing large-scale assessment frameworks in the country are largely normative and summative, and hence insufficient for a largely process-oriented CT education. Assessment remains rooted in the written examination mode, privileging knowledge items and skill demonstration rather than process-based understanding. For instance where the lesson analyses data in a railway timetable, classroom activities involve enjoyable trials of alternate data organisation, but assessments switch to numerical question answering. The development of reliable rubrics and metrics for CT, owned by the teaching community, poses a significant challenge, especially when the available educational resources are very limited.

In brief, our stand on these issues is to de-emphasise devices and computer programming during the first 8 years of school. We have tried to identify key components of CT in schools, covering both CT education as well as the possibilities of using CT in learning other subjects [19]. These include notions of scaling, iteration, data organisation, modelling, and algorithms, with reasoning about procedures providing the key element. These can be integrated into the mathematics curriculum during the first 8 years of school and can be achieved without access to devices or requiring additional curricular space. Further, the required reorientation of mathematics teachers may be easier to implement while teacher education is developed for CT as required. This will enrich mathematics education as well, as attested to by the Tamil Nadu experience. The conceptually based Assessment Frameworks advocated by NEP for mathematics education will also lend themselves to the process orientation required by CT education.

Such a stand necessarily ignores some dimensions of computing education. Moreover, for educational policy discussions underway in India today, these are issues of immediate large-scale importance. These considerations lead us to many questions related to computing education research, some of which are presented below.

- Does research show that children acquainted with programming from an earlier age make better programmers with sharper computing abilities?
- What are the implications of a curricular choice between teaching computing as a subject, and integrating it into other subjects, in elementary and middle schools?
- How do large-scale teacher education programmes address the challenges of a new discipline such as CT in schools and its assessment needs?
- Does the advent of AI/ML tools in education further alienate "unplugged" approaches to CT in schools?
- How does CT education fit in with considerations of 'holistic' education?

Indeed, this list of questions is not exhaustive, and more questions will arise at all levels of implementation of CT in the years ahead. However, answers to these fundamental questions are needed for clarity of direction to the teaching community.

5 In Conclusion

With the demand for computing in schools coming from many sections of society, educational policymakers tend to view it from the perspective of modernisation, especially in the Global South. There is a danger that this may translate to an exclusive emphasis on coding, with only lip service paid to the loftier aims of CT education. Not only to overcome this danger but also to utilise this historic opportunity to bring meaningful computing education in schools, we need to offer a process model that is implementable on a large scale, addressing low-resource educational environments. Our curricular experiments suggest that this is feasible, with some limitations. The demand for computing education at all stages of schooling calls for extensive research to provide guidelines on many aspects of CT curriculum and its place in school education.

References

1. ACM India: 4th conference on computational thinking in schools, 2022. https://event.india.acm.org/CTiS/2022/
2. ACM India: The CSpathshala curriculum (2016). https://cspathshala.org/curriculum/
3. Aoki, T.: Inspiriting the curriculum. In: Curriculum in a New Key: The Collected Works of Ted Aoki, pp. 357–365. Lawrence Eriblaum (2004)
4. Bell, T., Alexander, J., Freeman, I., Grimley, M.: Computer science unplugged: school students doing real computing without computers. N. Z. J. Appl. Comput. Inf. Technol. **13**, 20–29 (2009)
5. Chakraborty, P.: Computer, computer science, and computational thinking: relationship between the three concepts. Hum. Behav. Emerg. Technol. (2024). https://doi.org/10.1155/2024/5044787
6. Fullan, M.: The New Meaning of Educational Change. Teachers College Press, New York (2001)
7. Granone, F., Reikerås, E.K.L., Pollarolo, E., Kamola, M.: Critical thinking, problem-solving and computational thinking: Related but distinct? An analysis of similarities and differences based on an example of a play situation in an early childhood education setting. In: Paloma, F.G. (ed.) Teacher Training and Practice, chap. 4. IntechOpen, Rijeka (2023). https://doi.org/10.5772/intechopen.110795
8. Grover, S., Pea, R.: Computational thinking: a competency whose time has come, 2nd edition. In: Computer Science Education: Perspectives on Teaching and Learning in School, pp. 20–38. Bloomsbury (2023)
9. Kasturirangan, K.E.A.: National Education Policy (2020). https://www.education.gov.in/sites/upload_files/mhrd/files/NEP_Final_English_0.pdf
10. Krishanmoorthy, K., Pandiyan, M.: Integration of computational thinking in upper primary mathematics in Tamil Nadu, India. In: CTE-STEM 2021: APSCE International Conference on Computational Thinking and STEM Education, June 2021
11. Kwon, K., Jeon, M., Zhou, C., Kim, K., Brush, T.A.: Embodied learning for computational thinking in early primary education. J. Res. Technol. Educ. **56**(4), 410–430 (2022). https://doi.org/10.1080/15391523.2022.2158146
12. Macann, V., Yadav, A.: Culturally responsive-sustaining computational thinking: enactment in elementary classrooms. J. Comput. Sci. Integr. **6**(1), 7 (2023). https://doi.org/10.26716/jcsi.2023.12.22.51
13. National Council for Teacher Education: National Curriculum Framework for Teacher Education (2009). https://ncte.gov.in/website/PDF/NCFTE_2009.pdf
14. National Council of Educational Research and Training: National Curriculum Framework for School Education (2023). https://ncert.nic.in/pdf/NCFSE-2023-August_2023.pdf
15. Nevenglosky, E.: Barriers to effective curriculum implementation (2018). https://scholarworks.waldenu.edu/dissertations/5235
16. Oxfam India: India inequality report 2022, December 2022. https://www.oxfamindia.org/knowledgehub/workingpaper/india-inequality-report-2022-digital-divide

17. Palaparthi, P.: Computational thinking implementation in schools - an experience with rural welfare schools in India. In: Proceedings of Fifth APSCE International Conference on Computational Thinking and STEM Education Teachers Forum 2021, pp. 19–22, June 2021
18. Petta, L.D.: Computational thinking and unplugged activities: localization enabling learning. In: Proceedings of Fifth APSCE International Conference on Computational Thinking and STEM Education Teachers Forum 2021, pp. 23–25, June 2021
19. Ramanujam, R.: Computational thinking: the new buzz. At Right Angles **1**(13), 23–31 (2022)
20. Rittel, H., Webber, M.: Dilemmas in a general theory of planning. Policy Sci. **4**(2), 155–169 (1973). http://www.jstor.org/stable/4531523
21. Tamil Nadu SCERT: Tamil Nadu Curricular Framework 2017: A statement on Mathematics Curriculum, November 2017. https://tnprivateschools.com/Draft_Syllabus_Tamil_Nadu_2017/Position_Papers/Mathematics_PP.pdf
22. Tamil Nadu SCERT: Tamil nadu curricular syllabus for classes 1 to 10 (2018). https://www.trbtnpsc.com/2017/11/tamilnadu-new-draft-syllabus-2017.html
23. Ung, L.L., Labadin, J., Mohamad, F.S.: Computational thinking for teachers: development of a localised e-learning system. Comput. Educ. (2022). https://doi.org/10.1016/j.compedu.2021.104379

Why Teach About Binary Numbers?

Tim Bell and Henry Hickman

University of Canterbury, Christchurch, New Zealand
tim.bell@canterbury.ac.nz, henry.hickman@pg.canterbury.ac.nz

Abstract. This position paper considers the inclusion of binary number representation in school curricula. There can be resistance to this because it is seen as a mathematically advanced concept that isn't explicitly visible in digital technologies, and that there may be better things to spend curriculum time on. We argue that the key concepts are valuable for digitally literate students to understand, they exercise aspects of Computational Thinking, and that it is very easily introduced to young students. Binary digits (abbreviated by Claude Shannon to "bits") underpin all aspects of digital technology; importantly, they are the *digits* that make the technology *digital*, and therefore explain many of the benefits and limitations of digital devices. To reinforce this, we collect examples of where these digits are encountered in our digital society.

Keywords: Binary representation · Curriculum

1 Introduction

The topic of data representation appears in K-12 computing curricula in many countries [3]. It is usually based on the idea that computers represent all data using binary digits, and often expects students to demonstrate this understanding by working with binary representations, including converting numbers between decimal and binary. There are occasions where we are asked why students need to understand how binary numbers work; one keynote talk at a teacher conference even argued that teaching binary representation is "cruel" [4].

However, teachers have reported that if it is taught appropriately, many students find it very engaging and that primary school students pick it up quickly [2,8]. The main concepts can be grasped by a primary school student in 30 min or less[1]. Despite this ease of learning, and data representation, including

---

[1] The video at https://vimeo.com/342521353 shows students learning the key concepts quickly, and a related approach and full lesson plan can be found here: https://www.csunplugged.org/en/topics/binary-numbers/how-binary-digits-work/.

binary numbers, often being included in the curriculum, Falkner et al. found that many teachers weren't actually teaching the concept [3]. Perhaps some of the concerns around teaching "binary numbers" come from traditional teaching methods where the representation is expressed mathematically as powers of two and in the context of number bases, which involve notation that may be unfamiliar or uncomfortable for students and teachers. More engaging approaches involve constructivist teaching where students are challenged to represent numbers (and importantly, other information such as text and images) using only two symbols. The CS Unplugged approach achieves this by having them manipulate cards with 1, 2, 4, 8 and 16 dots on them respectively[2], and the CS Field Guide presents this at a level suitable for K12 school students[3]. Another approach is Curzon's "Locked-in syndrome" activity, which is based on how Jean-Dominique Bauby managed to author a whole book despite being completely paralysed except for being able to communicate two symbols: blinking and not blinking[4].

But even if it's easy to teach, this doesn't necessarily mean that we should teach it. In the following we explore two main reasons that there is considerable value in teaching this topic: (a) it exercises many aspects of Computational Thinking, and (b) it explains many of the limits and possibilities of technology.

The connection of binary data representation to Computational Thinking (CT) has been explained by Bell and Lodi [1], who show how an Unplugged binary activity can enable students to exercise and demonstrate key CT concepts (based on [6] and related definitions of CT). For example: abstraction, where two values (zero and one, yes and no, etc.) can represent a meaningful value; decomposition, where students are scaffolded to work out bit values one at a time; algorithms, such as discovering the process for adding one to a binary value; and logical thinking through reasoning about the uniqueness of the binary representation of a given value. Curzon's "Locked-in syndrome" activity (mentioned above), where students try to use just two symbols to communicate information, is actually titled "Searching to speak", and their discussion explores how a series of yes/no questions can be used to search a space, in this case, the possible characters in an alphabet. In fact, a view of the binary digits of a representation as a decision tree is covered in a classic Unplugged activity on information[5], and reveals that the greedy algorithm for binary conversion is just a binary search of the range of possible values.

In the remainder of this paper we explore some of the ways that the concept of binary representation enables and limits everyday technology.

---

[2] https://www.csunplugged.org/en/topics/binary-numbers/.
[3] https://www.csfieldguide.org.nz/en/chapters/data-representation/.
[4] https://teachinglondoncomputing.org/free-workshops/computational-thinking-searching-to-speak/.
[5] See page 40 of the Unplugged activity on information theory at https://classic.csunplugged.org/documents/activities/information-theory/unplugged-05-information_theory.pdf.

2 Exploring the Limits and Possibilities of Technology

Bits are the "atoms" of digital information. Claude Shannon introduced the idea of a *bit* as the fundamental unit of information, and he extended its importance from not just being a representation of data, but a way to compare the information content of, say, the random characters of a password with those in a very predictable sentence [7]. Everything stored digitally ultimately is represented using just two values, usually written as 0 and 1, but actually represented on devices in a variety of ways, such as using high and low voltages in memory, two directions of magnetism on hard disks, and frequencies of light in fibre-optic cable. It's not a secret code, but simply a very efficient and economical way to represent values on physical devices. We could have used decimal digits (0 to 9), but the precision required to do this would make computers slower and more expensive. This is quickly demonstrated by asking a class for the four values in the binary times table ($0 \times 0 = 0, 0 \times 1 = 0, 1 \times 0 = 0, 1 \times 1 = 1$). This shows how simple it is to work with binary values, and students often would prefer to use this new times table than the base-10 table that they have to memorise!

The concept of binary digits is behind why we use the term digital in "digital technology"—the power of digital representation is that everything is represented using binary values—photos, text, songs, scientific data, financial records, and in fact anything that is "digital". So a single hard disk can store music, seismic data, security camera footage; and a single network connection can be used to transmit videos, send emails, and view web pages.

Since all data is stored using binary digits, understanding them can be seen as equivalent to a chemist understanding atoms; much of what we observe in everyday chemical reactions doesn't need to be explained in terms of atoms, but to fully understand the process, getting down to that level gives an understanding of the full mechanism involved. Many people use computers successfully, and even have a career in computing, without ever knowing about binary representation, so the argument here isn't that it's essential, but that it's a teachable concept that underpins everything digital. Some students will be happy to be told "this value ranges from 0 to 1023", or "the subnet mask in a network is 255.255.255.0", but others will appreciate understanding why these numbers were chosen.

The use of binary digits to represent everything on digital devices also means that the limits—and power—of digital data are dictated by this representation. For example, for a long time text was represented on computers using the ASCII code, which allocated 7 bits per character; the 128 possible representations were sufficient to map to all the keys on a traditional computer keyboard, but excluded languages and cultures that use symbols other than the English alphabet. Character sets also demonstrate the power of this representation; if we compare an 8-bit code with a 16-bit code, the number of possible characters doesn't double, but increases from 256 to 65,536. Each extra bit doubles the possible range, and this exponential increase in power means that we don't need many bits to achieve extraordinary things.

Here are some specific examples of situations that affect people due to data being represented using binary values.

Media Quality. The quality of media is often measured in "bit depth"; 24-bit colour can represent images more accurately than the eye can perceive and 16-bit audio has more detail than the ear can detect. Recording studios often use 24-bit audio; how much more information compared with 16-bit audio? A naive answer might be that it stores 1.5 times the information, but it's actually 256 times more accurate. Questions like this hint at the logarithmic relationship between the number of bits and the information content, an underlying idea of Shannon's work on the entropy of information [7].

Computer Limits. Often we see "oddly specific" limits such as 255 characters in a text field, 256 members in a WhatsApp group, input values ranging from 0 to 1023, or a limit of 16 ports on a device. These are actually "nice round numbers", based on making full use of the range of binary values. A particularly common form of this is that a kilobyte is 1024 bytes, not 1000 bytes, or so we tell students who may not be ready to learn about the kibibyte. Memory addresses are represented using binary, so having memory in groups of 1000 would create awkward gaps in the range of values being represented. This also give insight into the difference between, say, 8-bit and 32-bit computers.

IP Addresses. On 24 September 2015 North America ran out of addresses for devices on the internet under the commonly used IPv4 system—these Internet Protocol addresses are the numeric value given to every device connected to the Internet (for example, 132.181.106.1). These values are 32-bit numbers, which allow for around 4 billion devices to be connected to the internet. When it was first deployed in 1983, the idea of having one device for everyone on the planet might have seemed extreme, but now many people have multiple devices, and we're having to move to a new version of the protocol, IPv6, which uses 128 bit numbers. This allows for up to 340,282,366,920,938,463,463,374,607,431,768,211,456 devices—surely we won't end up with that many? (Or will we? What if a person owns billions of nanodevices, each with their own address?) A related place that binary numbers are seen is in subnet masks, such as 255.255.255.0, which are 32-bit values written as 4 decimal numbers.

Time Overflow. GPS devices rely on accurate time measurements for calculations. The week is represented using just 10 bits, which means that once every 20 years the week number resets to zero. This last happened on 6 April 2019, and any devices working with GPS data needed to take account of this change. A new system is now being used that has increased to 13 bits instead of 10; this means the problem will occur every 157 years, and surely we'll have new GPS systems by then? (Or will we?!) There are many other time-related bugs looming due to the number of bits used to represent a date. Most are unlikely to be a problem if programmers are aware of them, but one that will be particularly important is the time rollover on 19 January 2038 for Unix-based systems (which includes vast numbers of devices that keep the internet running).

Choosing Integer Sizes. Many programming languages require the programmer to specify whether to use "long" or "short" integer values, which are typically 8, 16, 32 or 64 bits. Using a long value when a short one would do means that space is wasted in memory, which is an issue if you're storing millions or billions of values. But using a short value limits the range of numbers that can be stored.

One place where a programmer apparently chose too small a size for an integer is the original version of the software controlling the electrical system on a Boeing 787 Dreamliner, which overflows after 248 d. It appears that the problem was caused by using a 32-bit integer value[6].

Compression Fraud. In 2010 a man in Nelson, NZ, was found guilty of fraud by getting New Zealanders to invest $5.3 million in his data compression system that claimed it could losslessly compress any file to 7% of its original size[7]. A simple explanation based on the binary representation of files shows that this isn't possible and that the system must be fraudulent, but many people couldn't understand this and lost their life savings to him.

Cryptography. The Bitcoin currency is based on SHA-256 hashing; not only is the term "binary digit" embedded in the name, but understanding the level of security depends on understanding how many bits are used in the encryption protocol. For a long time the US government had a law that limited the use of encryption that used keys of more than 40 bits[8]. The significance of this law, and the effect it has on human rights, depends on the implications of the 40-bit limit. In these examples, bits are used as a measure of the strength of a crypto-system, and (in principle) each extra bit added to a key makes it twice as hard to attack.

Fig. 1. Rounding problems caused by binary representation in Scratch and Python

Weird Arithmetic. Many inaccuracies in arithmetic are explained by the binary representation. For example, in many programming languages (including Scratch and Python), if you wanted to subtract 20 cents from 30 cents by calculating 0.3 − 0.2, the result is 0.09999999999999998 (Fig. 1). If you subtract another 10 cents (0.1), the result won't be zero, and the program needs some attention to check if the amount owing is zero. This is essentially because the value 0.2 in binary is 0.00110011... recurring, and is rounded to a fixed number of digits, which is easily demonstrated using the Unplugged cards with fractional dots to the right of the units card.

---

[6] https://www.engadget.com/2015-05-01-boeing-787-dreamliner-software-bug.html.
[7] https://www.stuff.co.nz/nelson-mail/news/3892853/Whitley-found-guilty-of-fraud.
[8] https://www.govinfo.gov/content/pkg/FR-1996-11-19/pdf/96-29692.pdf.

Advanced Computer Science. Binary values and their limits come up more as students work with advanced topics in computer science. For example, the "textbook" implementation for a binary search to calculate a midpoint can cause an integer overflow [5]; a powerful data structure called the binomial heap depends on binary representation of the branch factors; and quantum computing extends the idea of bits to "qubits", which allow a superposition of the two states.

So many everyday activities—accessing the internet, using GPS, deciding the date—are vulnerable to problems with binary representations, and programmers especially need to be aware of these issues to write reliable software.

3 Conclusion

Teaching how binary representation works isn't about simply having students learn to convert numbers to binary, but for them to understand the patterns that affect digital devices in everyday life. It's important that teachers understand the purpose of teaching binary representation as a topic, so that they don't present it as some sort of necessary evil that appears in the curriculum, but use the idea to engage and empower students to exercise Computational Thinking to understand the atoms of their digital world better.

References

1. Bell, T., Lodi, M.: Constructing computational thinking without using computers. Constructivist Found. **14**(3), 342–351 (2019)
2. Duncan, C., Bell, T.: A pilot computer science and programming course for primary school students. In: Proceedings of the Workshop in Primary and Secondary Computing Education, pp. 39–48. WiPSCE '15. Association for Computing Machinery, New York (2015). https://doi.org/10.1145/2818314.2818328
3. Falkner, K., et al.: An international comparison of K-12 computer science education intended and enacted curricula. In: Proceedings of the 19th Koli Calling International Conference on Computing Education Research. Association for Computing Machinery, New York (2019). https://doi.org/10.1145/3364510.3364517, https://doi.org/10.1145/3364510.3364517
4. Mönig, J.: The music comes out of the piano. In: Keynote presentation at the European Conference on Technology-Enhanced Learning. ECTELFI '20, European Conference on Technology-Enhanced Learning (2020)
5. Pattis, R.E.: Textbook errors in binary searching. ACM SIGCSE Bull. **20**(1), 190–194 (1988)
6. Selby, C., Woollard, J.: Computational thinking: the developing definition (2013). https://eprints.soton.ac.uk/356481/
7. Shannon, C.E.: A mathematical theory of communication. Bell Syst. Tech. J. **27**(3), 379–423 (1948)
8. Takeharu Ishizuka, Tatsuya Horita, S.K.: Practical study on computer science unplugged for children aimed to considerer of application to the Japanese primary school education. In: Proceedings of the IFIP World Conference on Computers in Education, WCCE 2013, pp. 239–240. International Federation for Information Processing (2013)

Teaching Tangible Division Algorithms or Going from Concrete to Abstractions in Math Education by the Genetic Socratic Method

Juraj Hromkovič and Regula Lacher

ETH Zürich, Zürich, Switzerland
{juraj.hromkovic,regula.lacher}@inf.ethz.ch

Abstract. Education is about supporting humans in their growth, with a special focus on exploring their intellectual potential. Learning to act following a given (even complex) pattern is losing its educational value very fast, because all well described activities can be automized. Education therefore should focus on developing those cognitive process dimensions of pupils where technology cannot compete with humans. This means teaching how to describe and discover the world, how to verify own imaginations and models, how to think and how to design, analyze and evaluate new products of science and technology instead of learning the products of science and technology, and their applications.

In this article we claim that teaching numbers and fundamental arithmetic operations in schools starts on a too high abstract level resulting in learning algorithms of symbol manipulations without understanding the nature of the fundamental calculations.

In this paper we show that starting with the historical development of number representations (not with the decimal positional system) offers a natural, more understandable way for teaching mathematics in primary schools. We show, that going consequently from concrete to abstract empowers pupils to be able to design own representations of numbers and rediscover the execution of arithmetic operations on their own. We take the operation of division of integers to exemplary illustrate how a successful process of rediscovery of arithmetic algorithms can be designed.

Keywords: teaching fundamental arithmetic operations · critical thinking · number representations · division with coins · abstraction · genetic Socrates method · constructivism · algorithmic symbol manipulation

1 Introduction

"The brains of people should not be stuffed with facts, names, and formulae.
To know all this, it is not necessary to have passed any school.
The real purpose of education is to teach people to think."[1]

---

[1] Albert Einstein.

Automation has been always the part of human culture that made and makes our civilization more and more efficient. Since ever humans acquire knowledge and use it to develop procedures (algorithms) for different purposes. The original principle of automation was that many people could successfully execute such procedures without understanding why they work properly, i.e., without having the knowledge of their investigators. The oldest archeological artefacts documenting algorithms as mathematically described exact procedures are about 4000 years old. In this sense computer science as discipline about automated information processing has been always an integral part of human culture. Today we are living in the era of information technology that enables to automize all activities we understand to some extent and execute them faster and more reliable than humans. The 200 years old model of schools striving to educate experts able to correctly act following a complex pattern needs to be updated because the educational value of acting by following given procedure descriptions (does not matter how complex they are) is decreasing fast. Since education is about supporting humans to grow (especially to explore their intellectual potential), nowadays one has to force the development of those dimensions of pupils (creativity, phantasy, critical thinking) in which technology cannot compete with humans.

We claim that the current style of teaching numbers and basic arithmetic operations in primary schools does not fit our requirements formulated above. Introducing calculation starts on an abstraction level that is too high. The abstract, decimal positional number representation is assumed as given and it is ignored that the written algorithms for multiplication and division are products of thousands of years of finetuning to minimize the amount of work for their execution as well minimize the space of their symbolic (abstract) execution. And of course, the optimization of these two aspects and searching for an appropriate number representation impacted each other and was done simultaneously. Teaching these final products as one without being familiar with the process of their development aggravates considerably the understanding of basic arithmetic. Each abstraction arises as a generalization of experiences with concrete, and this experience with concrete is the only well understandable way for introducing abstractions and working with them (Jean Piaget [1, 2]). It is a folklore that many institutions educating teachers decided not to aim to teach the execution of arithmetic operations because this typically results in learning the corresponding symbol manipulations without understanding why these algorithms work. But this is a wrong decision because avoiding abstractions restricts the intellectual growth of pupils a lot. Without learning to create and verify abstract models (descriptions) of objects and processes we essentially restrict the ability of learners to approach more complex topics (a more involved discussion can be found in Hromkovič, Lacher, Marty: Mathematik entdecken und entwickeln, to appear (in German) [3–7]). This is the reason why we start to teach calculation with numbers by developing various and more concrete representations of numbers and designing algorithms for the basic arithmetic operations in these less abstract number representations. In this paper we show part of our design of a teaching sequence devoted to the understandable development of an abstract representation of the division procedure. The main didactic strategy is based on splitting the learning path in such small steps (sequences of questions, tasks, and activities [8–10]), that pupils are able to make the necessary discoveries for solving particular steps (problems) to a high extent on their own.

To learn how to describe objects (numbers, for instance) and calculation processes in abstract ways is more important than to learn executing concrete calculation algorithms. The ability to abstract is the key in the processes of understanding and shaping the world around us.

This paper is structured as follows. In the second chapter we shortly discuss the development of number representations with the focus on so called coin-representations. In Sect. 3, we show how to develop a very well understandable approach for a division algorithm based on two simple modules – exchanges a big coin for several small coins, and equally partitioning a set of equally-valued coins (splitting a pirate treasure). In Sect. 4 we discuss how to move from this transparent tangible algorithm to an understandable script form (written execution). In conclusion we discuss the importance of our approach of teaching to calculate.

2 Developing Number Representations

The history of developing number representations is at least 68.000 years old. The first attempts resulted in the unary number representations which is suitable for small numbers only. Nevertheless, historically this was one of the greatest breakthroughs in human history enabling the storage of information outside of the human brain. The true digital revolution started more than 5000 years ago with the development of the first scripts in Mesopotamia and then in Egypt as the consequences of the first Big-Data crises in human history. Using alphabets in order to represent shortly (concisely) information as sequences of symbols (today called data as digital representations of information) and store them on media has changed the management of and the living in the old cultures more essentially than the IT does recently. For the first time in the history of mankind, humans were able to store information externally outside of their brains on some medias. This allowed humans to store information more objectively and for an unrestricted amount of time, to communicate information to arbitrary long distances, to build databases for information processing, and to do business with information.

From the very beginning the number representations of most of the old cultures have been based on addition. One has chosen basic values of the number system used (called coins here), for instance the decimal ones (1, 10, 100, 1000, ...) in Egypt, the hexadecimal ones (1, 60, 3600, ...) in Mesopotamia, or the vigesimal system (1, 5, 20, 400, 8000, ...) by Maya in South America. The origin of the Roman number representation is the same by taking coins of values 1, 5, 10, 50, 100, 500, and 1000. In all these cultures a number was expressed by a collection of coins (basic values) in such a way as the sum of the values of the coins in the collection corresponds to the value of the number. Obviously based on this principle, one can express a larger number by many different coin collections. To avoid ambiguity in their number representations all old cultures have agreed that the collection with the smallest number of coins expressing a given number is its right representation. For the coins (basic values of the number systems) used by old cultures this representation strategy guarantees the unambiguous representation of all numbers.

From the didactic point of view one of the main advantages of the coin-number representations is that one can work with numbers in a physical representation as collections of coins. This number representation is so transparent that pupils in the second class of the primary school do not have any problem to count and to sum numbers up to 1 million. Addition is especially easy to execute. One takes two numbers in their coin representations as coin collections and unite these two collections into one collection. The sum of the values of the coins in this unified collection offers already the right value of the addition. The pupils are now asked to minimize the number of coins by exchanging smaller coins for equal value of larger coins. Training this exchange of coins without changing the value of a coin collection is the only pre-knowledge we require to learn the tangible division algorithm presented in the next chapter.

3 Division in Coin Representations of Numbers

Teaching the division, one starts with the unary number representation, which in tangible representation means with a collection of equal-valued coins. The fair distribution of a coin collection is running in rounds and in each round each party gets one coin. The transparent procedure is illustrated in Fig. 1.

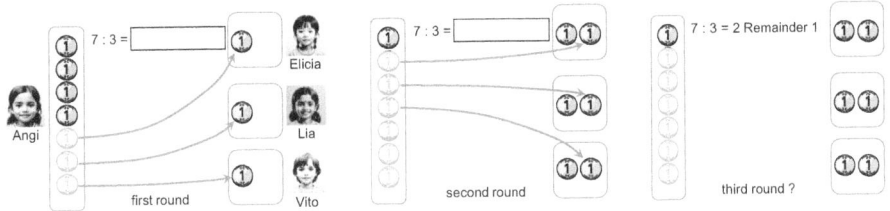

Fig. 1. Distributing coins to three people.

We see that the fair partitioning of coins in this way can be mastered by pupils in a few minutes and one can easily learn to write the result of the division following the common convention. This simple procedure is the basic module for the general division algorithm. This module has to be applied so many times as the number of positions of the positional number representation of the dividend. Below (Figs. 2, 3, 4 and 5) we see that one begins with distributing the coins of the largest value on the corresponding number of parties given by the divisor. If some rest coins cannot be distributed, one exchanges them for the coins of the next smaller value. This repetition of distribution of equal-valued coins and exchange for smaller valued coins is transparently executed in any number system, not only in the presented decimal system. We encourage teachers to apply this tangible division strategy also for other number systems, because this contributes a lot to understanding of the process of division.

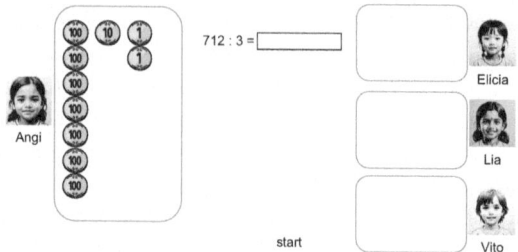

Fig. 2. Distributing coins of different values; start.

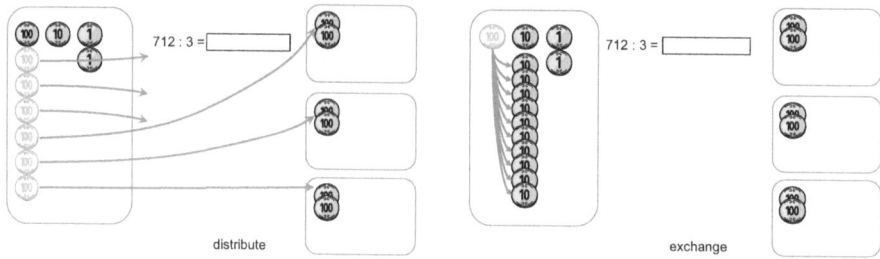

Fig. 3. Distributing coins of different values; first distribution and exchange.

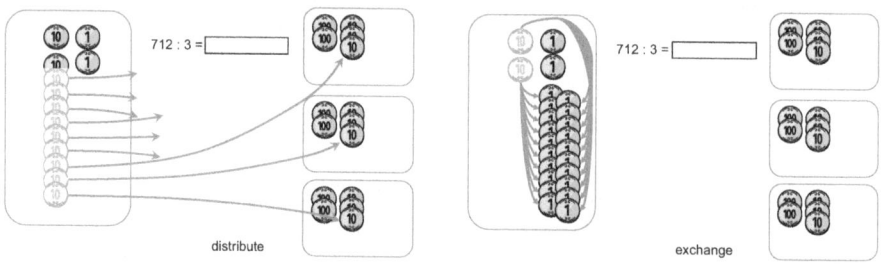

Fig. 4. Distributing coins of different values; second distribution and exchange.

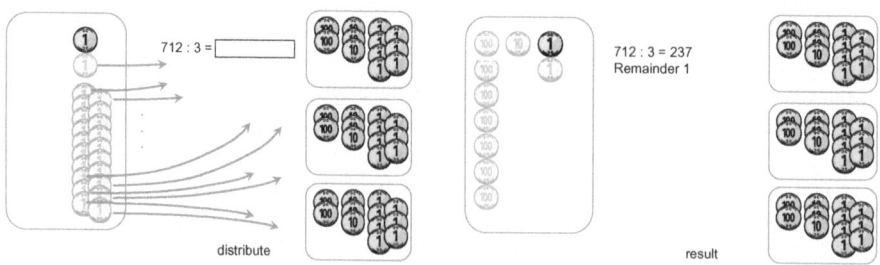

Fig. 5. Distributing coins of different values; last distribution and result.

4 Division as a Symbol Manipulation in Digital Number Representations

As we have seen above, one can learn to understand division, and divide numbers by small divisors very quickly in a transparent and tangible way. The main difficulty in teaching division is not in understanding its execution, this can be mastered in one lesson. The hardest task is to learn to execute the division in an abstract representation as an algorithm manipulating symbols with which the numbers are represented. We do not recommend to directly go from the presented tangible division execution in coin representations in one step to a written representation of the division procedure although the written representation is only an exact description of the tangible representation introduced. To learn to divide in a written form we developed a sequence of steps moving carefully from concrete manipulation to its written form. Here we only show the most important step of moving from concrete to abstract. The idea is to avoid the optimization that strives to execute division in the smallest possible place with the smallest number of symbol manipulations. In the representation below we use two columns. The left column always involves the coins that still have to be distributed (labelled "to distribute"), and the right column contains the already distributed coins (labelled "distributed"). The time of the execution is running top down and so the sum of the coins in each row correspond to the value of the dividend. In this transparent description pupils can reconstruct any particular step of division in the concrete and tangible execution (Figs. 6, 7, 8, 9, 10 and 11).

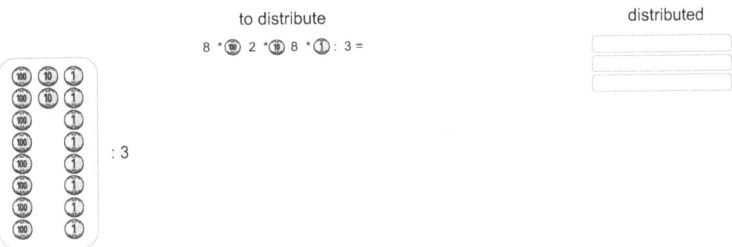

Fig. 6. Semi-written representation; Start.

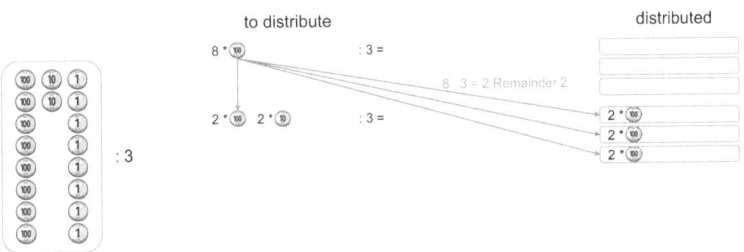

Fig. 7. Semi-written representation; first distribution.

First, one starts with divisions using only a small number of coins during the division execution. In this way one can work only with symbols for coins drawing the corresponding coins. Later one can continue as in the illustration above where we already use digits

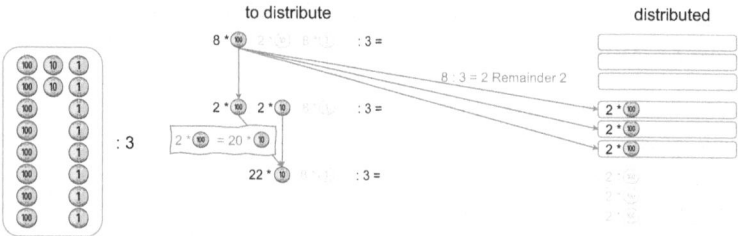

Fig. 8. Semi-written representation; first exchange.

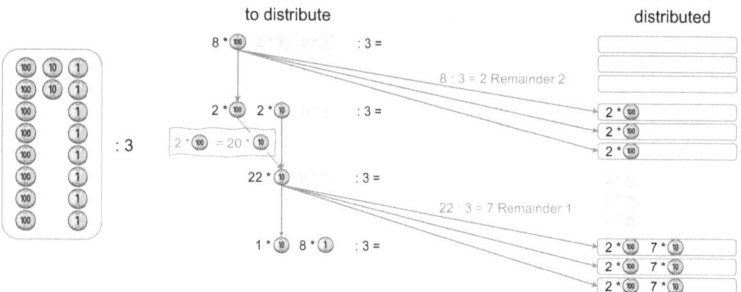

Fig. 9. Semi-written representation; second distribution.

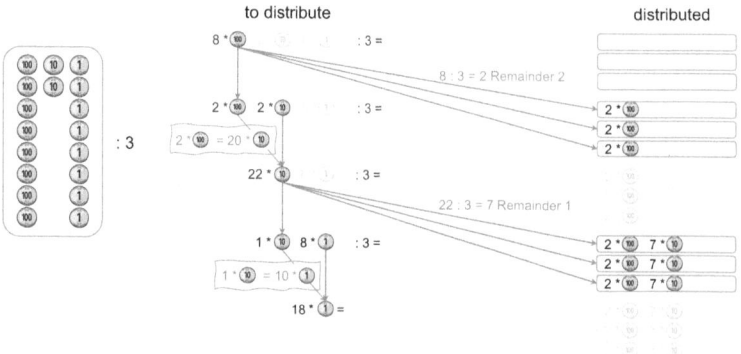

Fig. 10. Semi-written representation; second exchange.

in order to write the number of coins used instead of drawing the corresponding number of coins. Note that using cards with digits, one can execute the division in the illustration above in a tangible way and therefore keep the relation between the execution and its description transparent.

Some more modules and steps are needed to develop the written division for large divisors, but this is not the issue for the first years of the primary school.

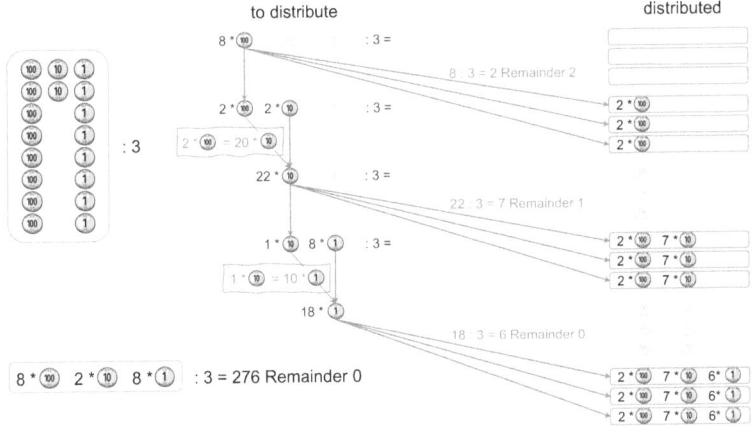

Fig. 11. Semi-written representation; last distribution including the result.

5 Conclusion

For kids as novices with a very low experience in abstraction there is no other way than starting with concrete and learn to develop the abstraction as a generalization of their experience with concrete by active handling (Piaget [1, 2]). The hardest part of teaching the execution of arithmetic operations is not in understanding their meaning and in their execution in a transparent and tangible representation, but in their common, size-optimized written representation. If we omit to work in concrete representations, we risk that pupils will learn to calculate without understanding the meaning of what they do. For preparing the lessons and writing textbooks we recommend using the genetic Socratic method [8–10].

The Motto is: "Do not focus to teach the products of science and technology (facts, theorems, models, methods, etc.) and to work with them but teach the processes of their discoveries and their development."

Especially this means to study the genesis of the products of science and to also recognize the failures that helped to achieve breakthroughs. According to this one can divide the processes of inventions in such small steps (sequence of questions, tasks, challenges, and activities [8–10]) that pupils can discover particular steps with high probability on their own. We illustrated this approach by teaching division. Observe that after discovering the distribution of equal-valued coins in the first lesson presented above, the pupils to a high extent are able to participate in the development of the tangible division algorithm which makes the acquired knowledge much more sustainable. The same is true if pupils are involved in the development of the written description of the already mastered tangible division procedure.

References

1. Piaget, J. (1926). La représentation du monde chez l'enfant. F.Alcan, Paris. Translation to German: Bernard, L. (1978). Das Weltbild des Kindes, Klett-Cotta, Stuttgart

2. Piaget, J., Harel, I.: The Psychology of Intelligence. Harward University Press, Cambridge (1950)
3. Hromkovič, H., Lacher, R., Marty, J.: Mathematik entdecken und entwickeln, Praxisbuch 1, Von Zeichen zu Ziffern, Schubi Lernmedien AG, Schaffhausen (to appear)
4. Hromkovič, H., Lacher, R., Marty, J.: Mathematik entdecken und entwickeln, Praxisbuch 2, Von Zahlendarstellungen zum Plus, Schubi Lernmedien AG, Schaffhausen (to appear)
5. Hromkovič, H., Lacher, R., Marty, J.: Mathematik entdecken und entwickeln, Praxisbuch 3, Vom Plus zum Mal, Schubi Lernmedien AG, Schaffhausen (to appear)
6. Hromkovič, H., Lacher, R., Marty, J.: Mathematik entdecken und entwickeln, Praxisbuch 4, Vom Plus zum Minus, Schubi Lernmedien AG, Schaffhausen (to appear)
7. Hromkovič, H., Lacher, R., Marty, J.: Mathematik entdecken und entwickeln, Praxisbuch 5, Vom Minus zum Teilen, Schubi Lernmedien AG, Schaffhausen (to appear)
8. Delic, H., Senad, B.: Socratic method as an approach to teaching. Euro. Res. **111**, Is. 10, 511–517 (2016). https://doi.org/10.13187/er.2016.111.511
9. Wagenschein, M.: Zum Problem des Genetischen Lehrens, Zeitschrift für Pädagogik (1966)
10. Wittmann, E.: Grundfragen des Mathematikunterrichts (6th edition), Vieweg, Braunschweig/Wiesbaden (1981). https://doi.org/10.1007/978-3-322-91539-9

Computational Thinking
and Interdisciplinary Instruction

Mathematical *Versus* Computational Thinking with a Computer in the Background

Maciej M. Sysło[✉]

Warsaw School of Computer Science, Warsaw, Poland
syslo@ii.uni.wroc.pl

Abstract. In this article, we focus on the use of mental tools of computational thinking (CT) to solve selected problems in school mathematics. We suggest how to expand and enrich some of the traditional school mathematics topics through the use of CT, and as a result, obtain solutions that use and are supported by the power of informatics as a discipline and computers as computational tools. We show how addressing the computational complexity of problems can help also with developing mathematical thinking (MT). Although the problems are mathematical in nature, they do not appear in teaching mathematics in Poland according to the actual core curriculum, but they do appear in selected informatics classes (All references to the core curriculum in this paper are to the National Core Curriculum approved by the Ministry of National Education of Poland in 2017/2018). In the first part of the work, we briefly refer to computational thinking, mathematical thinking and general problem-solving strategies.

Keywords: Computational Thinking · Mathematical thinking · Algorithmics

1 Introduction, from History

The use of calculators and computers in mathematics has been a topic of discussion at conferences held in Poland since the early 1980s. During the ICMI Symposium (on Mathematics Instruction) at the ICM Congress in Warsaw in 1983, Zofia Krygowska, a leading mathematics educator in Poland, and Bernard Cornu, a computer scientist with a mathematical background made some very important observations.

Z. Krygowska (Krygowska, 1986) argued: "Mathematical concepts are operational in nature. The young mind is wide open to the operationalization of mathematical content, because it is often interested in "how it is done"? and "how it is constructed"? rather than in "what this is?". This is particularly conducive to introducing the student to algorithmization processes. Coding and schematizing activities find a particularly natural basis here, even in children starting school."

B. Cornu (Cornu, 1986) claimed that: "informatics changes not only the methods of teaching but also the contents of teaching. Calculators can be used in many ways in the classroom. [...]. But, above all, informatics is a new pedagogical tool which makes possible the achievement of a new approach to mathematics: active mathematics, experimental mathematics, visualization, elaboration of conjectures, simulation. [...] calculators modify relations between the teaching, the student, and the contents of teaching."

Erik de Corte, the invited speaker at the National Conference "Informatyka w Szkole" in 2001, justified that a major reason underlying the failure of educational computing is that the computer is mainly introduced as an add-on to an existing unchanged classroom setting which leads only to reproduction and preservation of the status quo. It is a wrong assumption that computers will evoke by themselves productive learning, whereas productive educational computer applications require to be embedded in teaching-learning environments, see also Papert, 1980.

Two important conclusions from the above presentations guided our computer scientists' proposals and activities in the years that followed, with respect to computers in informatics, mathematics, and other fields of education. Unfortunately, with only moderate success, especially in teaching and learning mathematics, as can be seen in the actual core mathematics curriculum, which does not include any algorithm, computer or technology, but only "students perform calculations using mathematical tables and calculators". A salvation for the students are informatics classes, where they have the opportunity to encounter mathematical tasks and issues, enriched with informatics elements that are integrated with the tasks. In Sect. 3, we look at some examples that appeared in the author's earlier publications. Here, however, we clearly refer to the use of computational thinking and its tools, and we also draw attention to their place in the spiral development of informatics and mathematics competences, which constitute the foundation for the structure of the core curriculum for both subjects. In the proposed methods of implementing individual topics, we suggest many students' activities that, in our understanding, involve a computer in the background (Sysło, 2023). As we used to suggest to them: think computationally (in this case also mathematically) before you start programming and using a computer.

In the next Section, we briefly introduce the basics of computational thinking, mathematical thinking, and problem solving, which are the main area of students' activity when implementing the topics presented in Sect. 3.

2 Definitions

In this Section we discuss definitions of three concepts used in this paper: computational thinking (CT), mathematical thinking (MT), and problem solving. These terms often appear in the literature in various contexts, often together, with emphasis on their different meanings and scopes. For this reason, it is difficult to provide a clear description that fits different situations, as in this paper. Since our main focus is on the role of CT in mathematical thinking during problem solving, we will refer both terms related to thinking to the problem solving process.

2.1 Computational Thinking

As we wrote in (Sysło, 2023), since 1997 for the next 15 years in Poland, all core informatics curricula confirmed Denning's opinion that: Computational thinking has a long history within computer science. Known in the 1950s and 1960s as "algorithmic thinking," (Denning, 2009). In the informatics curriculum for high schools approved in 1997, one can read in the section "Algorithmics and Programming" that "the school is to provide conditions for students to acquire the following competences:

1. Define a problem situation, including data [*abstraction*], the goal and the results.
2. Formulate a plan for solving the problem – separate sub-problems [*pattern recognition, decomposition*] and indicate connections between them.
3. Choose a way to solve the problem:
   - design an algorithm [*algorithmic thinking*].
   - use an existing program or program a solution method in a selected programming language [*implementation, programming*].
4. Analyze the correctness of the algorithm and its implementation [*debugging*], and assess its complexity [*evaluation*], test the program [*testing*].
5. Complex projects solve in a team [*collaboration*].
6. Choose and solve problems from various school subjects [*generalizations*]."

The above list of competencies is similar to the operational definition of CT. We have also inserted some mental tools of CT [*in italic*] that constitute another definition of CT. Thus, CT has a long tradition in our informatics education as operationally defined set of mental tools used in the process of solving problems to learn informatics concepts and apply methods (algorithms). Therefore, the following definition of CT fits our approach (Wing, 2014): *Computational thinking is the thought processes involved in formulating a problem and expressing its solution(s) in such a way that a computer – human or machine – can effectively carry out.*

It should be obvious that the above steps, some or all, of solving problems with the use of mental tools of CT do not always require the use of a computer, although it can be used at any time if the person solving the problems so decides, see (Sysło, 2023) about informatics lessons with a computer in the background.

Because CT did not emerge directly from our approach to teaching informatics through problem solving, we did not also teach CT directly, and this is often the case today. We rather teach how to discover, develop, and use CT in solving problems from various areas of education. Similarly, as we suggest not to "teach Python" but to "teach programming using Python". Therefore, we avoid to use the terms "CT education", "teaching CT", "CT classes" and similar, as used by many authors. CT is an approach and a set of mental tools used in solving problems associated with learning informatics concepts and methods.

Since many problems solved in our informatics classes are related to mathematics, our informatics classes also expand mathematical thinking by applying CT in solving mathematical problems. Moreover, as we show in the next Section, this is a significant extension of the scope of problem solving to include issues related to the effectiveness (computational complexity) of the obtained solutions [*to be effectively carried out*, (Wing, 2014)], also in relation to their practical implementations and applications.

2.2 Mathematical Thinking

Mathematical thinking (MT) has a much longer tradition than computational thinking (CT), see (Schoenfeld, 1995) and (Stacey, 2006). Despite this, MT and CT are closely related today and many mental tools of CT are also found in MT, such as logical thinking and abstract thinking. On the other hand, some CT mental tools, such as recursive thinking and logarithmic thinking (see Sect. 4), are heavily based on MT, but CT adds very important aspects of computational complexity to them, as we illustrate in Sect. 3.

Although CT and MT are of different origin, both are used in the process of solving problems, although CT usually accompanies MT but MT not always appears with CT. Mental tools, such as: problem decomposition, abstraction, algorithmic design, iteration, and generalization are common for CT and MT, although the scope of their use may vary and be different. There is a growing interest in and use of mental and computational tools of CT in mathematics and in solving mathematical problems. Heuristic strategy is common for both.

2.3 Problem Solving

> The mathematician's main reason for existence is to solve problems.
> [P. Halmos]

The definition of algorithmic thinking cited in Sect. 2.1 can be interpreted as the process of building a solution of a problem in the form of an algorithm which then can be programed and run on a computer. The description of this process naturally includes concepts such as data structures, algorithm, and program, because a computer is mainly used to execute programs, and each program is a record of an algorithm that processes data. In a sense, algorithmic thinking leads to a computer solution to the problem. However, there are many problems that we want to solve without using a computer or we are only interested in providing a solution without running it on a computer, see (Sysło, 2023).

The field of solving problems, especially mathematical ones, is much older than computer problem solving. George Pólya is considered a pioneer in the field of problem solving. In his epoch-making book *How to solve it* (Pólya, 1957) he defined the basic steps in the problem-solving process and also characterized an approach to problem-solving called **heuristics**. His approach can be applied to problems in almost any field, including informatics problems (see Sect. 4.4).

Pólya's general problem solving method consists of the following four steps:

P1. Understanding the problem
P2. Devising a plan
P3. Carrying out the plan
P4. Looking back

We refer the reader to Pólya's book, where the author comments in detail on these steps in the problem-solving process and provides an extensive "short dictionary of heuristics". There are versions of these 4 steps, e.g. expanded to 6, as well as other formulations (Schoenfeld, 1995). We may also try to match the 6 steps of algorithmic thinking (Sect. 2.1) with these 4 steps.

The problems we solve in Sect. 3 meet the requirements that Pólya sets for problems: a problem is not an exercise of some mathematical skill or procedure that is already known; a problem which involves problem solving is "a task for which the solution method is not known in advance". In each of our cases, students are faced with a problem situation that is completely new to them. Developing solutions to these problems there, however, does not fully fall under either of the two problem-solving patterns presented in this chapter. This is an important comment: the problems are different and rarely the process of solving them can be summarized in 4 or 6 steps.

3 Solving Some Mathematical Problems with CT

In this Section we present a number of topics, which are a part of informatics education in the schools in Poland and we show here how they could also contribute to mathematics education.

Donald Knuth, in his work on algorithmic and mathematical thinking (Knuth, 1985), based on the analysis of selected pages in nine academic mathematics textbooks, noticed that one of the missing elements in the thinking of mathematicians is the computational complexity of algorithms. Professional mathematicians do not pay much attention to the efficiency of mathematical calculations. Presumably for this reason, the authors of the core mathematics curriculum in Poland did not include the efficiency of calculations in the curriculum.

As computer power increases, it may seem that striving for more and more efficient methods of solving problems no longer makes sense. However, the statement of Ralph Gomory (IBM Fellow, 1959–1989) still holds true, see also Sect. 3.5:

> ... one way to make the hardware go faster
> is to make the hardware work with less steps ...

As we will show with some mathematical problems, the use of CT mental tools in designing and developing their solutions leads to deeper insight into the problem, discovering related concepts, as well as much more efficient ways of solving the problems, both for humans and computers. The process of developing their solutions illustrates also integration of both subjects – mathematics and informatics. Moreover, it shows also how informatics and CT tools, when embedded in the process, change not only methods of learning and teaching but also what students learn (Cornu, 1986).

3.1 The Problem with Many Triangles

We begin with an example of a problem which could be used to illustrate a difference between two of its solutions, one given in a mathematics class and another developed during an informatics lesson. The problem is a simplified version of the problem from the Informatics Olympiad (Sysło, Kwiatkowska, 2006).

Problem. We are given a set A consisting of a large number of integers, e.g. 50 million. Verify if each triple of numbers from A can be the lengths of sides of a triangle. *Note*, the assumption about the size of A is intended to make the student understand that the entire set A cannot be saved in the computer's memory (it was in 1993).

According to the core mathematics curriculum for grades 4–6 (age 11–13): "the student constructs a triangle with given three sides and determines the possibility of building a triangle based on the *triangle inequality*". Therefore, for the possible lengths of the sides of a triangle x, y, z she/he checks three inequalities:

$$x + y > z; \quad x + z > y; \quad y + z > x$$

Solution I (in a mathematics class). The solution is very "simple" – a student suggests to check the triangle inequality for all possible triples of elements from A. If there are n integers in A, then Cn^3 operations (additions and comparisons), where C is a constant, are performed in this case. A very impractical solution.

Solution II (in an informatics class). We challenge students if they could reduce the number of inequalities they have to check. We guide their intuition by asking them to draw a situation in which a triangle cannot be formed from three segments. They discover that in such a case two shorter segments together are shorter than the longest of the three segments. Hence, they conclude that it is enough to check one inequality, whether the two shortest segments are in total longer than the longest one. After a short discussion the students conclude that if the triangle inequality must hold for all triples in A, then we have to test the triangle inequality for only the two smallest integers, min_1 and min_2, and the largest integer max in A.

Then, the next challenge is how to find these three numbers in this huge set? Luckily, according to the core informatics curriculum for grades 4–6 (age 11–13): "a student: formulates and writes commands in the form of algorithms for [...] finding an element in the set, unordered or ordered, the smallest and the largest". Therefore during the informatics classes the students learn how to find the smallest and the largest elements in a set by applying a **linear search** and in fact we do not need to sort all numbers in A to find these three numbers – the following statements are sufficient:

```
min₁:= ∞; min₂:= ∞; max:= - ∞;
while there is new data x in A do begin
    if x < min₂ then
        if x < min₁ then begin min₂:= min₁; min₁:= x end
        else min₂:= x
    if max < x then max:= x
end
```

Finally only one inequality $min_1 + min_2 > max$ must be verified. If it is satisfied then the triangle condition holds also for all triples of elements from A.

Conclusion. The computer solution is very efficient, at most three comparisons and two assignments are performed in each step. Thus, the total complexity is bounded by $C'n$ operations, where C' is a constant and n is the number of elements in A. Hence, Solution II is linear in the size of the input and moreover it may be applied to data set A of any size input from any source. It is not only very efficient solution but also it has a very simple implementation – the program is short and the whole solution is elegant. A computer/informatics introduces a new dimension to solving mathematical problems.

3.2 Searching and Sorting

Finding the smallest or largest (or best) element in a set of elements of a given type is one of the most frequently used "activities" in everyday life. It is therefore important for students to realize that they can solve these problems in the most efficient way by using linear search.

Linear search used to solve the triangle problem is one of the most important algorithms used to solve problems on unordered sets of elements. Let us mention here Hugo Steinhaus's beautiful argument presented in his excellent book (Steinhaus, 1989) that finding the *max* or *min* of n unordered numbers requires at least $n - 1$ comparisons. It looks like this: let's imagine that we are looking for the *max* using a traditional tennis tournament. If Iga wins this tournament, what can be said about all the remaining $n - 1$ players? Each of them had lost at least once, so at least $n - 1$ matches were played. Hence it follows that the algorithm for finding *max/min* among n unordered numbers that performs $n - 1$ comparisons is optimal.

In (Sysło, Kwiatkowska, 2014a) we discuss several other problems and show how linear search can be applied to many other situations. These problems illustrate the spiral approach and the mental tool of generalization as students solve the problems and must adapt the linear search to new situations, and also reductive thinking which depends on reduction of a problem to be solved to use the solution of the problem which is known how to solve. Some of the solutions the students arrive at are also optimal (finding the second player in a tournament or finding *min* and *max* simultaneously). They also learn that using optimal linear search does not necessarily guarantee an optimal algorithm for some other problems, such as selection sort.

3.3 Searching in an Ordered Set – Find a Word, Guess a Number

Problem. Imagine an encyclopedia of fishes that has 1000 pages. Find the page that has information about catfish by checking the least number of pages. While searching for the right page, students discover **binary search**, which is dividing the remaining pages into two equal parts and eliminating the one that does not contain the catfish until only one page remains where the catfish should be.

This is an old-fashioned problem since today students rarely look at paper books, they prefer electronic books, which are equipped with a search mechanism. However, a two-person game guessing a number hidden in a given interval is very popular among young people and may serve the same purpose. We ask students to play several rounds of this game in pairs and to fill in Table 1. One student chooses an interval and gives it to the other student. The first student then chooses a number in that interval, but does not tell it to the second student, who has to guess that number. He guesses by asking whether it is a given number and receives one of the answers: yes, smaller or larger. They note in the table the number **gol** – after how many questions the second student guessed the hidden number. We also placed in the last two columns number called **log** and the base 2 logarithm of the interval length.

The table below shows the results of the game from the moment when the student guessing the number aimed at the middle of the remaining interval to search. Additionally, regardless of the hidden number, the log number is equal to the number of times the size of the interval is divided by 2 to get 1 (when dividing an odd number by 2 we take "the larger half"), for example: 30, 15, 8, 4, 2, 1. The numbers in the last column can be added later. In each game, the numbers in the last three columns should be close together and gol $\leq$ log.

Table 1. The results of two-person guessing game.

Interval	Interval size	Hidden number	**gol** – number of questions asked	**log**	⌈$\log_2$(Interval size)⌉
[1, 225]	225	225	8	8	8
[51, 180]	130	100	6	8	8
[500, 1000]	500	900	9	9	9

Two related issues emerged in the discussion with students.

- Let's assume that we have a paper telephone book in which the entries are arranged alphabetically by the names of the telephone owners. To find Mr. Kowalski's phone number, we use binary search and among 1000 pages we find the one with Kowalski's phone number after visiting at most 10 pages. And how many pages do we have to look through if we want to find the name of the owner of the phone number 1234567? All 1000! This best illustrates the power of both, ordering information and the binary search algorithm.
- The students modified the binary search by claiming that when they want to find a word in an encyclopedia starting with one of the first letters of the alphabet, they usually try to find this word on the first pages – such a strategy is called **interpolation search** and is described in (Sysło, 2016).

3.4 Binary Representation; Size of a Number in a Computer

One of the first steps in the problem-solving process is the analysis of data and the selection of their proper representation. In particular, knowledge of how numbers are represented in a computer allows us to take into account the properties of this representation in the solving process, as well as its impact on this process, and especially the efficiency of this process. In this section we deal with the binary representation of numbers (integers), and in the following sections we will use its properties in the design and analysis of selected algorithms.

We again refer here to the core informatics curriculum, now for grades 7–8 (age 14–15), where we can read that: "a student presents ways of representing logical values, natural numbers (binary system), characters (ASCII codes) and texts in a computer". Therefore during the informatics classes the students learn how to find the **binary representation** of integers.

n	q	r
14	7	0
7	3	1
3	1	1
1	0	1

For a given non-negative integer n, such a representation is generated by successively dividing n and obtaining quotients by 2. They divide n by 2 and take the remainder r (0 or 1) as the least significant digit of the representation. Then apply this procedure to the quotient q and continue until the quotient is zero. For $n = 14$ we get $(1110)_2$ as in the table.

Then we ask students, how many binary digits has a decimal integer n in its binary representation or equivalently, how much space in the computer memory is needed for storing n. The answer to this question is very important since it is related to solutions of many theoretical and algorithmic problems. In (Sysło, Kwiatkowska, 2014c) we use simple mathematics to show how to obtain the answer to this question, it equals exactly $\lceil \log_2(n+1) \rceil$. So we may assume that the integer n occupies about $\log_2 n$ bits in the computer's memory, and this number is often taken as **the size of n in the computer**.

Further discussion leads students to the conclusion that a binary search on an interval of size n is equivalent to finding the binary representation of n – in both cases we start with n and divide by 2 until we reach 1. Putting these three arguments together we conclude that the number of steps in a binary search over an interval of size n equals approximately $\log_2 n$. *Note*: in this discussion we do not refer to the definition of logarithm, which is only introduced in high school. Finally we conclude with the **algorithmic definition of $\log_2 n$**:

The base 2 logarithm of n – $\log_2 n$ – is equal to the number of steps in which dividing n and subsequent quotients in dividing by 2 leads to 1.

We demonstrate in (Sysło, Kwiatkowska, 2014c) how important is this interpretation of logarithm and logarithmic thinking for designing and evaluating solutions to many problems, see also the next Sections.

3.5 Fast Exponentiation

First, we refer to the core mathematics curriculum for grades 7–8 (age 14–15): "a student multiplies and divides powers with positive integer exponents; multiplies powers with different bases and the same exponents; raises power to power".

In mathematics lessons, students practice exponentiation usually for very small exponents, single digits. However, in many real-world applications, such as compound interest and cryptography RSA, exponents can be very large. It is worth making students aware that even a supercomputer will not be able to help in such cases. Let the exponent consists of 32 digits: $n = 12345678901234567890123456789012$ and we have an access to a supercomputer that performs 10^{18} op/s. How long will it take this computer to calculate the power with this exponent if the exponentiation is performed as successive multiplications ($n - 1$ of them for exponent n)? Students successively divide n by: 10^{18}, 60, 60, 24, 365 and get 391479. Therefore, it will take the supercomputer $3*10^6$ years to calculate this "small" power. However, exponents in RSA cryptography have several hundred digits – how is it possible that we receive messages encrypted with this system in a fraction of a second?

To guide students to faster exponentiation, we suggest they decompose the exponent into smaller exponents. First, they quite easily notice the decomposition of an even number $n = 2k$ into two factors: 2 and k. However, they generally have trouble with odd exponents, so we suggest to derive the even exponent from the odd exponent using the formula $n = 2k + 1$. In the next step, students apply these rules to calculate, for example, x^{14} and they get:

$$x^{14} = \left(x^{2*7}\right) = \left(x^{7*2}\right) = \left(x^7\right)^2; x^7 = x^{6+1} = x^6 * x; x^6 = \left(x^3\right)^2; x^3 = x^{2+1} = x^2 * x; x^2 = x * x$$

Therefore, to calculate x^{14}, the multiplications are performed from right to left and instead of 13 multiplications using the school method, only 5 multiplications (squaring is one multiplication) are performed. All operations can be put in one expression:

$$x^{14} = \left(\left(x^2 * x\right)^2 * x\right)^2$$

This expression suggests its recursive implementation:

```
Power(x,n)                              {xⁿ}
  if n = 1 then Power:=x
  else if n - even then
           Power:=Power(x,n/2)^2        {xⁿ = (xⁿ/²)²}
       else Power:=Power(x,n-1)*x       {xⁿ = (xⁿ⁻¹)x}
```

Then the teacher analyzes with students what is the relationship between the number of multiplications in the above algorithm and the value of the exponent n. Here it will be useful to remind students a binary representation of numbers. They should observe that in the binary representation of integer numbers, even numbers have 0 and odd numbers have 1 in the least significant position. Hence, students should draw the following conclusion: each position in the binary expansion of the exponent, except the first one, corresponds to one multiplication and an additional multiplication if there is 1 in the position. Therefore, the number of multiplications performed in the above algorithm is not greater than twice the length of the binary expansion of the exponent minus 1. For $n = 14 = (1110)_2$, 5 multiplications are performed (3 positions plus 2 ones), which is smaller than 2 times 3.

number of decimal digits in n	2 x binary length of n
100	664
200	1330
300	1994
500	3322

Let us now estimate how many multiplication will perform the above algorithm for n consisting of 32 digits. The exponent n occupies 104 binary positions therefore the algorithm will perform at most 208 multiplications. It is a tremendous achievement in efficiency – we get the result after 208 multiplications instead of waiting 10^6 years! This is usually a shock for students! The table shows that calculating x^n for n with hundreds of digits takes only a few thousands of multiplications.

3.6 Euclid's Algorithm

We refer again to the core mathematics curriculum for grades 4–6 (age 11–13), which in the section "Operations on Natural Numbers" states that: "a student:

- performs division with the remainder of natural numbers;
- decomposes two-digit numbers into prime factors;
- finds the greatest common divisor (GCD) in situations no more difficult than types GCD(600, 72), GCD(140, 567), GCD(10000, 48), GCD(910, 2016) and determines the least common multiple of two natural numbers using the method of factorization;
- determines the result of dividing the number a by the number b and writing down the number a in the form: $a = b*q + r$."

The suggested GCD calculation for two numbers a and b using their prime factorization teaches factorization and shows the benefits of prime factorization. However, for slightly larger numbers, factorization may cause problems for students and discourage them from using this method of calculating GCD. For example, calculating GCD(4807,7163) can quickly become boring for students, even though both numbers are the products of three two-digit numbers.

A more serious objection to the proposed method results from the complexity status of the factorization problem – there are no known fast factorization methods (operating in polynomial time with respect to the number of binary digits of the number being factorized). However we can easily overcome this difficulty referring to the last statement from the curriculum above which says that students can "determine the result of dividing the number a by the number b and writing down the number a in the form: $a = b*q + r$". When we additionally impose the condition that the remainder r is smaller than the divisor b, i.e. $0 \leq r < b$, then q and r are uniquely determined for given a and b.

The relation $a = b*q + r$ between a and b leads to the following conclusions which can be turned into Euclid's algorithm for calculating GCD(b, a):

1. If $r = 0$, then b divides a, i.e. GCD(b, a) = b.
2. If $r \neq 0$, then we have $r = a - q*b$, i.e. every number that divides a and b also divides r, especially the largest such number, hence GCD(b, a) = GCD(r, b).

The last equality leads to the recursive implementation of Euclid's algorithm:

a	b	r
34	21	13
21	13	8
13	8	5
8	5	3
5	3	2
3	2	1
2	1	0

```
GCD(b,a)
    if b = 0 then GCD = a
    else GCD = GCD(a mod b,b)
```

Euclid described his algorithm 300 years BC,[1] he was not interested in its efficiency. However a teacher with her/his students can analyze its behavior and conclude about its complexity. First, students apply the algorithm to different pairs of integers and enter the results of each step of the calculation in the table as above. The table shows the results for the pair $a = 34$ and $b = 21$. Then we suggest students to compare the numbers in columns 1 and 3 in the same row. They should notice that in the same row, the number in the third column is at least twice smaller than the number in the first column. This fact, that for each row the remainder r from dividing a by b is not greater than $a/2$, can be shown very rigorously (Sysło, Kwiatkowska, 2014c). Therefore, in the sequence of remainders generated by Euclid's algorithm, each number is at least two times smaller than the number that appears two positions earlier. It reminds a sequence generated by a binary search except a sequence of the numbers generated in Euclid's algorithm could be twice longer. Hence we may estimate that Euclid's algorithm, finds GCD(b, a) in at most $2\log_2 n$ steps, hence it's efficiency is similar to the efficiency of binary search. Hence, one may conclude that Euclid was very close to invent logarithm, almost 2000 years before John Napier did it in 1614!

We leave to students a challenging question, for which numbers a and b does Euclid's algorithm makes the largest number of steps. We used such a pair in the table.

4 Computational Thinking – Various Ways of Thinking

Computational thinking has many faces in every field, just look at the work (Wing, 2006). In mathematics, due to its close relationship with informatics, it finds applications and can be found in almost every subject area, in relation to almost every problem. This also applies to school mathematics. Below, we briefly supplement the above examples with remarks about selected varieties of thinking that complement the most frequently mentioned mental tools of CT.

4.1 Recursion – Thinking Recursively

Jeannette Wing wrote "Computational thinking is thinking recursively" (Wing, 2006). Over 20 years earlier, Seymour Papert, referring to the Logo environment, wrote (Papert, 1980; p. 74): "Of all ideas I have introduced to children, recursion[2] stands out as the one idea that is particularly able to evoke an exited response. I think this is partly because the idea of going on forever touches on every child's fantasies and partly because recursion itself has roots in popular culture."

---

[1] An interesting 2-person game can be used to introduce Euclid's algorithm in unplugged manner (Cole, Davie, 1969).

[2] Papert used a tail recursion in his book.

In Sects. 3.5 and 3.6 we showed how recursion can be used to design and develop solutions to some problems which occur in school mathematics. In both cases, recursion involves creating a solution by reducing to the same problem, but of a smaller size, and predicting when it will end. This approach can be considered as reductive thinking, but it is important to understand the mechanism of recursive calls, so the proper name is **recursive thinking**. The difficulty for students is to understand how a given problem can be solved by referring to the same problem?! Without going into too much detail, they should be convinced by ensuring them that the computer "knows" how to perform successive recursive calls up to a stopping condition. Many people treat recursion as a kind of a loop. Indeed, recursion is a type of repetition, but its proper model is the copy model – at each stage of the recursive call, the same module is copied and executed.

In general, recursion is a tool for problem-solving by decomposition of a problem into subproblems of the same kind – one has to specify two decisions: how to decompose a problem into subproblems (e,g. by divide and conquer method) and then how to compose the solutions of the subproblems into the solution of the problem. Calculating Fibonacci numbers recursively is an example of such an approach: two previous numbers are independently calculated recursively and then their values are added.

Recursion should not be taught as a separate subject, rather it is a way of thinking, used in various situations. In some cases it is redundant (e.g. when adding a sequence of numbers), it can be an alternative to iteration, but we mainly focus on its properties as a concept that has computational power when designing solutions to problems and running such solutions on a computer (Sysło, Kwiatkowska, 2014b).

4.2 Generalizations – Reductive Thinking

In Sects. 3.1–2 we used **reductive thinking** (Armoni et al., 2005) in solving selected problems that involve finding elements in unordered sets. This is a natural approach in the spiral development of students' skills, where they have to take what they have already learned and apply to more general problem situations. The ability to use reduction is characteristic of CT, see more details in (Sysło, Kwiatkowska, 2014b).

4.3 Logarithm – Think Logarithmically!

In this paper we touch on many problems whose common feature is a close connection with logarithm, and more generally – with logarithmic thinking. The suggested approach to solving selected problems, which in general involves dividing data/problem into halves, leads to solutions whose efficiency is logarithmic with respect to the computer size of the data. This is a significant achievement from the point of view of the algorithmic and complexity properties of the solutions. On the other hand, this type of algorithmic thinking contributes to a better understanding of the concept of logarithm and its role in the design of practical algorithms and computations, see more details in (Sysło, Kwiatkowska, 2014c).

4.4 Mathematical Thinking

Computational thinking covers various aspects of problem solving in many areas of mathematics. A large area of computer calculations are engineering calculations, which are performed on approximate numbers and using approximation algorithms. In school mathematics, for example, students calculate the approximate value of the square root of 2. The approximate results of computer calculations have to do with the fact that real numbers are saved in the computer in an approximate way, the function values are approximate because the computer performs only four operations exactly, and for many problems, no effective algorithms are known, so approximation algorithms are used. The approach to designing approximate calculations can be called **approximation thinking**. An important factor in such calculations is the accuracy of calculations.

Due to George Pólya, **heuristics** is an approach to solving mathematical problems. However, in informatics, this concept is defined slightly differently. This is an approach used to find an approximate solution when other methods fail to find an exact solution. The purpose of heuristics is to find a solution in a reasonable time that is good enough from the point of view of the assumed goal. We can therefore talk about **heuristic thinking** if the goal is to design and obtain a solution to the problem that meets the user's expectations. Heuristic thinking is a very powerful tool in the hands of students, used to discover and create a solution, even when they are beginners in the area of the problem. Greedy methods as heuristics are very popular, especially in the area of optimization where the goal is to find the best possible solution, e.g. an optimal map coloring, finding a shortest path, finding a travelling salesman tour.

Acknowledgements. The author thanks the reviewers for their valuable suggestions that improved the presentation.

References

1. Armoni, M., Gal-Ezer, J., Tirosh, D.: Solving problems reductively. J. Educ. Comput. Res. **32**(2), 113–129 (2005)
2. Cole, A.J., Davie, A.J.T.: A game based on the Euclidean algorithm and a winning strategy for it. Math. Gaz. **53**(386), 354–357 (1969)
3. Cornu, B.: The role of calculators in teaching mathematics for all (in Polish). Dydaktyka Matematyki, 83–104 (1986)
4. Denning, P.J.: (2009), The profession of IT beyond computational thinking. Comm. ACM **52**(6), 28–30 (2009)
5. Knuth, D.E.: Algorithmic thinking and mathematical thinking, the american math. Monthly **92**, 170–181 (1985)
6. Krygowska Z.: Components of mathematical activities which should play a significant role in mathematics education for all (n Polish). Dydaktyka Matematyki 25–41(1986)

7. Papert S.: Mindstorms. Children, Computers, and Powerful Ideas, Basic Books (1980)
8. Pólya, G.: How to Solve It. Princeton University Press (1957)
9. Schoenfeld, A.H.: Mathematical Problem Solving. Academia, New York (1995)
10. Stacey, K.: What is mathematical thinking and why is it important? Manuscript (2006)
11. Steinhaus H.: Mathematical Snapshots, G.E. Stechert Press 1939; 3rd edition, Oxford University, Press, New York 1969; WSiP, Warszawa 1989 (1989)
12. Sysło M.M.: Algorithms (in Polish), Helion (2016)
13. Sysło M.M., Kwiatkowska A.B.: Contribution of informatics education to mathematics education in schools, In: ISSEP 2006, LNCS 4226, pp. 209–219 (2006).https://doi.org/10.1007/11915355_20
14. Sysło, M.M., Kwiatkowska A.B., Learning mathematics supported by computational thinking. Constructionism Creativity, 258–268 (2014a)
15. Sysło M.M., Kwiatkowska A.B.: Introducing students to recursion: a multi-facet and multi-tool approach, ISSEP 2014, LNCS 8730, pp. 124–137 (2014b).https://doi.org/10.1007/978-3-319-09958-3_12
16. Sysło, M.M., Kwiatkowska, A.B.: Think logarithmically! KEYCIT **2014**, 371–380 (2014c)
17. Sysło M.M.: Education with a Computer in the Background, ISSEP 2023, Local Proceedings, 89–101 (2023)
18. Wing, J.M.: Computational thinking. Comm. ACM **49**(3), 33–35 (2006)
19. Wing J.M.: Computational Thinking Benefits Society (2014). http://socialissues.cs.toronto.edu/index.html%3Fp=279.html

Computational Thinking Based STEM Art Exhibits

Jay Thakkar[✉] and Manish Jain

Center for Creative Learning, Indian Institute of Technology Gandhinagar, Gujarat 382355, India
j.thakkar@iitgn.ac.in

Abstract. Making large-scale STEM exhibits can be a very engaging group activity for students across all ages. Apart from giving them a sense of accomplishment from completing the mammoth task of exhibit making, it also inspires them to think about the underlying algorithm that generated the design. In this paper, we describe exhibit designs based on pixel art using materials such as dice, bindis, Rubik's cubes, strings, tessellation tiles, sticky notes, push-pins etc. We also share our experience and learnings from making 25+ different large scale portraits with students from elementary school to undergraduates. Affordable raw-materials and open-source tools make the designs accessible for use by educators in their schools.

Keywords: Pixel Art · Big Exhibit · String Art · Computational Thinking

1 Implementation

An image in a computer is represented using pixels and their intensity. If it is a gray-scale image, the intensity or brightness will range from pixel values 0 to 255. The 0 value represents the darkest shade (black) and 255 represents the brightest shade (white). There are many pixel-art based works [1–4] which promote integration of art and computational thinking. We have created an open-source tool [6] which algorithmically maps each pixel value with a side of the dice's face or size of a bindi (circle).

Consider a black colored dice with white dots on its faces representing numbers 1–6. In the portrait design with dice, the brightness of the pixel is mapped with six sides (one to six dots) of the dice. The pixel of an input image with value 0 (black color) will be mapped to the face of the dice having one dot (face with maximum black area). Similarly, a pixel with 255 value (white color) will be mapped to the six-dot face (face with minimum black area). All the other ranges for brightness of the pixel are linearly mapped with the faces of dice. A similar algorithm is used to create pixel-art with different sizes of bindi. The bright pixels are mapped with smaller size bindi and dark pixels are mapped with bigger size bindi. The open-source tool [7] designed by us takes the input image and creates a template to paste the bindis. The number in the template (generated by the software) represents the size of the bindi (Fig. 1).

Another category of portraits is based on string art. The perimeter of a circle is divided into equidistant points and an image is constructed by weaving strings between these points. Figure 3 shows the 10 feet diameter exhibit of the IIT Gandhinagar logo

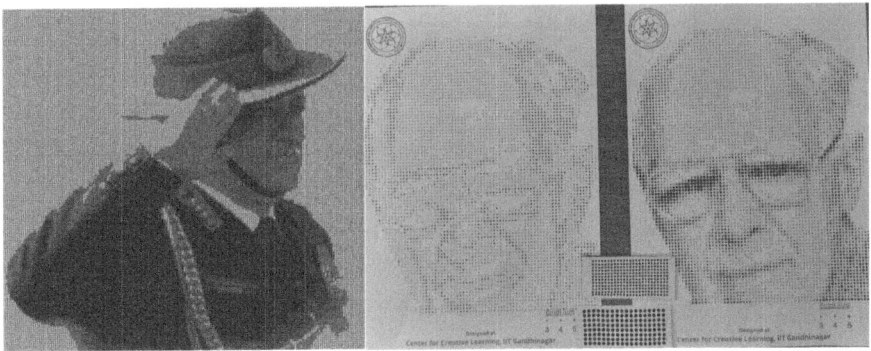

Fig. 1. Portrait of Late General Bipin Ravat using 42,000 dice & Portrait of Martin Gardner using total 1300 bindis of 3 different sizes

using 10 km long thread (4000 straight lines) weaved among 300 points on a circle. 200 undergraduates at IIT Gandhinagar made this in 6 h as part of the foundation program 2022. CCL created an open-source tool [8] to generate string weaving patterns to construct images of any text. For a given alphabet, the tool generates a sequence to weave strings onto the points on the circle. Consider the pattern generation of the letter C in the text CMSC. The perimeter of the circle is divided into 150 points (decided by trial-and-error for various letters for optimum result) and labeled 0 to 149. The starting point is 0 and there are 149 straight lines from point 0. The line is chosen based on a greedy approach - (i) First, find the set of pixels for every line (ii) Calculate the brightness of each line; this is calculated by taking the average of brightness of every pixel in the line (the set of lines in step (i)). (iii) Choose the darkest line among the 149 sets. Now, connect the string between point 0 and the destination point derived from the algorithm. Repeat the process (remove the selected line from the set while scanning the darkest line) for drawing 1500 straight lines. Figure 2 shows the output of the mentioned algorithm for the text CMSC. The tool outputs a spreadsheet with 1500 sequences of numbers to weave a letter. It takes 90 min to weave 1500 straight lines in a circle with 150 nails.

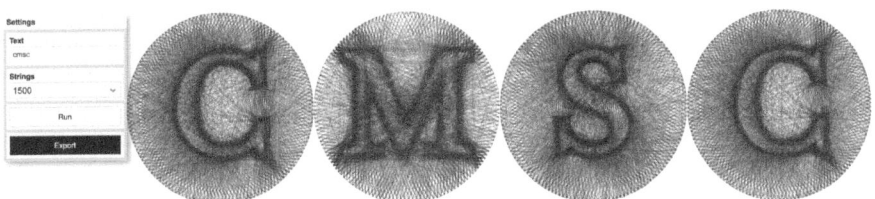

Fig. 2. "CMSC" string art using 1500 straight lines and 150 points on a circle

The Center has created more than 25 such portraits and exhibits based on pixel art. Discussing the implementation of every design is beyond the scope of the paper. A glimpse of the work is given in the next section.

Fig. 3. 10 feet IITGN logo using 4000 straight lines and 300 points on a circle

2 Impact

In the year 2022, we worked with 75,000 middle-school students of 746 Kasturba Gandhi Balika Vidyalaya in Uttar Pradesh. We attempted to evaluate the engagement of portrait making based on Bindi Art in online mode. Initially, the students felt it challenging to arrange bindis individually in bulk. But then, teachers modified it to a group activity and the whole class created the set of portraits using bindis. Some girls even generated their own templates from the open-source tool created by the Center. The artwork was then followed by discussions on how the computer screen creates a gray-scale image and coloured image using only 3 colors (RGB). By zooming in on the computer/mobile screen, we illustrated how a compact and huge number of simple arrangements of 3 light sources can create such a realistic image.

All these exhibits are experimented with a wide spectrum of participants. Some of the examples are: portrait of Dr. A.P.J. Abdul Kalam (see Fig. 4) using 32,400 push-pins to 180+ middle school children from various government schools in Gujarat; portrait of Vikram Sarabhai and Asima Chatterjee (see Fig. 4) with string art by master trainers of Agastya Foundation; portrait of Dr. C.V. Raman (see Fig. 4) using 625 Rubik's cubes by 400+ high school students. In an online STEM program, the Center published a module on Bindi Art and received 150+ projects (see Fig. 4) from children and teachers showcasing the potential of the activity as a CS-Unplugged activity in regular classrooms. Before starting the instructions of exhibit making, the students are given a flavor of how the computer algorithms have been used in creating the logo/portrait. For example, the greedy algorithm [4] of choosing a sequence of a chord in a circle to create any image, mapping pixels of a colored image to the Rubik's cube color palette and image dithering etc. Considering the scale of the big exhibits and the efforts involved, the goal is to give sense-of-achievements to the freshers along with letting them realize the underlying design algorithms.

At school level, the focus is on making a small scale art-piece i.e. making A3 paper size portraits using bindis, shading the pixel-blocks using pencils (labeled from 1 to 6 based on darkness). The objective is to let them appreciate the simple principle behind

Fig. 4. Colored portraits using push-pins, Rubik's cubes and String art (Color Figure Online)

how all images are being displayed on screen by just varying the darkness of the pixel. As a part of a weekly online program with 75,000 children from middle schools in India, the center conducted a session from CS Unplugged Module on Image Representation [4]. Along with illustration of how image is represented, the students were also asked questions during and after the session on binary representation, type of shades given number of bits etc. The session concluded with a week-long project of making portraits of various national figures of their own choice and the template PDF [9] were provided online. The template includes A2 sized portrait (divided into 4 A4 sheets) templates with more than 10,000 pixels per portrait (Fig. 6). We were surprised to receive submissions [10] from 465 schools out of 1100 schools across 4 states (Uttar Pradesh, Gujarat, Rajasthan and Madhya Pradesh) in India (Fig. 5).

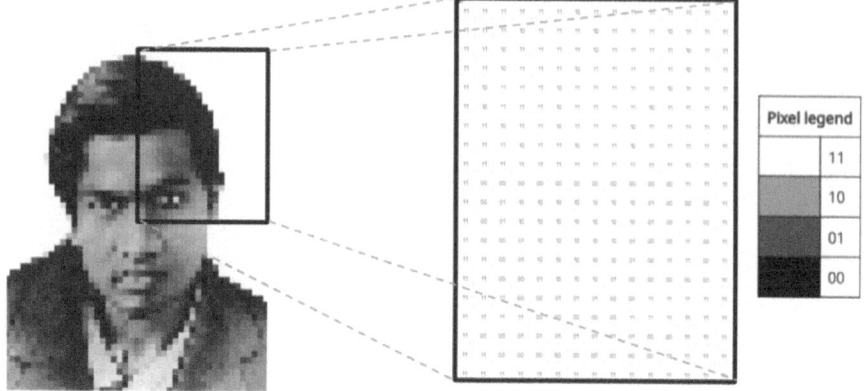

Fig. 5. Online session on CS Unplugged: Image Representation with Middle School Girls

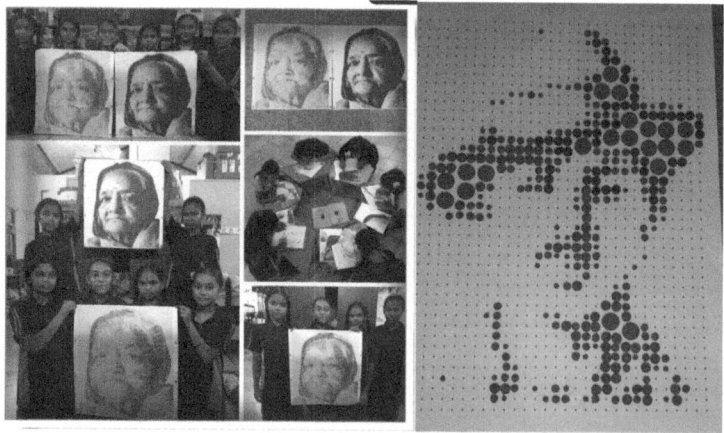

Fig. 6. A2 sized Portrait of Kasturba Mohandas Gandhi & bindi portrait made by students in the online STEM program by CCL IIT Gandhinagar

3 Conclusion

Integrating Art and an unplugged way of introducing computational thinking makes the portrait-making activity engaging. Various designs of exhibit making are executed by the center to show the feasibility of such computational thinking based large-scale projects on various extracurricular school occasions.

Acknowledgments. Thanks to the entire team of Center for Creative Learning, IIT Gandhinagar for execution of various exhibit making events.

References

1. Bosch, R.: Constructing domino portraits. Tribute Mathemagician 251–256 (2004)

2. Bosch, R., Adrianne, H.: Continuous line drawings via the traveling salesman problem. Oper. Res. Lett. **32**(4), 302–303 (2004)
3. Birsak, M, et al.: String art: towards computational fabrication of string images. In: Computer Graphics Forum, vol. 37, No. 2, pp. 263–274 (2018)
4. Bell, T., Michael, L.: Computational Thinking through unplugged activities
5. Bosch, R.: Opt Art: From Mathematical Optimization to Visual Design. Princeton University Press, (2019)
6. Tool to create portrait using Dice. https://tinyurl.com/DicePortraitCCL
7. Tool to create portrait using Bindis. https://ccl.iitgn.ac.in/bindiart
8. Tool to create String Art for a given text. http://ccl.iitgn.ac.in/stringart
9. Templates to make portraits. https://tinyurl.com/PixelArtTemplatesKGBV
10. Submissions on pixel art session. https://tinyurl.com/KGBVPixelArtSubmissions

BeLLE: Detecting National Differences in Computational Thinking and Computer Science Through an International Challenge

Heidi Kaarto[1(✉)], Javier Bilbao[2], Arnold Pears[3], Valentina Dagienė[4], Janica Kilpi[1], Marika Parviainen[1], Zsuzsa Pluhár[5], Yasemin Gülbahar[6], and Mikko-Jussi Laakso[1]

[1] Turku Research Institute for Learning Analytics, University of Turku, Turku, Finland
{heidi.kaarto,mhparv,milaak}@utu.fi
[2] Applied Mathematics Department, University of the Basque Country (UPV/EHU), Bilbao, Spain
javier.bilbao@ehu.eus
[3] KTH Royal Institute of Technology, Stockholm, Sweden
pears@kth.se
[4] Vilnius University, Institute of Educational Sciences, Vilnius, Lithuania
valentina.dagiene@mif.vu.lt
[5] Eötvös Loránd University, Budapest, Hungary
pluharzs@inf.elte.hu
[6] Teachers College, Columbia University, New York, USA

Abstract. The Bebras challenge is an international initiative to engage school pupils with computer science and computational thinking via an annual challenge designed by computer science experts and educators. BeLLE is an international consortium focusing on international comparisons of standard challenge task banks. The consortium uses the ViLLE platform to manage the challenge as it offers a digital exercise-based learning environment with comprehensive learning analytics. The goal is to learn more about students' knowledge in computer science and computational thinking in order to provide information for curriculum development and other educational planning and research. BeLLE started in 2021 with a pilots in Hungary and India. After the successful pilot, the consortium expanded substantially in the following year, many multiple choice questions were transformed into interactive tasks, and the process of organizing the challenge in ViLLE was refined. In this paper, we present some results of the challenge conducted within the BeLLE consortium in 2022. Over 90,000 students (47 % girls, 53 % boys) participated in the challenge in total through BeLLE. The results show that most students across countries and age groups get less than half of the maximum scores. The difference between girls and boys become apparent in Hungary and Lithuania: in the two oldest age groups (14–16 and 16–19 years old) boys score higher than girls. The time spent on the

- challenge is often 30 to 40 min with a difference between girls and boys in Hungary and Lithuania: boys use either less or more time than girls.

Keywords: computer science · computational thinking · challenge · gender difference

1 Introduction

Computational thinking (CT) is a problem-solving approach that involves breaking down complex problems into smaller, more manageable sub-problems and designing effective and efficient solutions using computational concepts and techniques. It is a fundamental competence in computer science (CS) and has increasingly gained recognition as an essential competence for everybody in the digital age. At its core, CT involves analyzing problems and formulating algorithms to solve them. It requires a combination of logical reasoning and the ability to think in multiple levels of abstractions [2]. By applying these concepts and techniques, people can approach problems in a structured and systematic way to ensure better problem-solving and solutions [8].

One notable initiative that promotes CT among students globally is the Bebras challenge [3], aiming to introduce students to CT and CS concepts in a fun and engaging way. It provides a series of short tasks (problems or puzzles) as an annual challenge where students are able to try solving the tasks as many times as needed in a specified time frame.

The Bebras challenge is designed to foster CT skills, logical reasoning, and problem-solving abilities. It encourages students to think computationally and develop a mindset that can be applied not only to CS but also to various other disciplines and real-life situations. The challenge offers different age-specific categories, ensuring that students from different grade levels can participate and benefit from the experience.

Participating in the Bebras challenge has numerous educational benefits. It nurtures an interest in CS and technology. By exposing students to CT at an early age, it helps demystify CS and encourages students to pursue further studies and careers in the field. It also promotes diversity and inclusivity, as the challenge is open to students of all backgrounds and abilities. Incorporating CT and initiatives like the Bebras challenge into education is vital for preparing students for the digital future. It equips them with the necessary skills to navigate an increasingly technology-driven society. By integrating CT into the curriculum, educators can empower students to become active problem solvers, critical thinkers and lifelong learners [4].

In education, one of the most important ideas is to get students from different backgrounds to the same level for a fulfilling future. With this goal in mind, the ViLLE learning environment was developed in Finland. ViLLE is an UNESCO awarded [11] exercise-based digital learning environment that offers automatically assessed exercises with immediate feedback for students while teachers can focus on teaching instead of grading. ViLLE offers the teachers comprehensive learning analytics from the first lesson in ViLLE to the last lessons and exams [7].

ViLLE's ready-made learning materials in mathematics have been proven to increase students' knowledge in short term [5] and long term [6]. In addition to learning materials used year round, ViLLE also has a built-in assessment tool which has been used to assess students' mathematical skills [1]. The assessment tool has been adapted to Bebras challenge in 2021 for the first time: the research consortium combining ViLLE and Bebras challenge is known as BeLLE. The goal is to learn more about students' knowledge in CS and CT in order to provide information for curriculum development and other educational planning and research.

The BeLLE consortium started with a pilots in Hungary and India followed by the official Bebras challenge in Hungary. The pilots were successful: teachers and students of all ages found the system to be easy and enjoyable to use. The results of the Hungarian challenge were also analyzed; the scores from girls and boys in age groups IV, V and VI (corresponding to ages from 12 to 18) show significant differences with boys getting better scores than girls [9].

In this study, we are looking at the Bebras challenge in 2022 when the official Bebras challenge within this research consortium was organised in six countries (Cuba, Dominican Republic, Hungary, Lithuania, Spain and Türkiye). The goal of this study is to explore the students' performance in the challenge across multiple countries and the differences between girls and boys' results. This could give us relevant information about the needs demanded for the curricula of different countries and then improve the implementation and development of CS and CT during primary and secondary schools. In addition, it can also indicate needs in the field of training for teachers in CT.

2 Research Questions and Methodology

The goal of this study is to investigate the students' performance in the Bebras challenge across multiple countries and see what kind of differences there are between girls' and boys' results. This main goal could help us to find keys for improving the implementation of CS and CT in the different curricula of the countries. The research questions are as follows.

1. How did students perform in the Bebras challenge in 2022?
2. What kind of differences were there between girls and boys in the challenge?

While analyzing the results, we need to remember the influence of differences in translations, culture, background knowledge, the educational process and the motivation of the students since we have multiple countries participating in the challenge. ViLLE and the Lithuanian challenge system used the same interactive tasks (created in Bebras Lodge, the Lithuanian interactive task creation system) to make sure that the interactivity of tasks would be as similar as possible.

The data is collected in ViLLE and the Lithuanian challenge system, and therefore, before the challenge started, we decided on the attributes we wanted to collect and made sure both systems collected them as similarly as possible. The final data includes all submission by the students since students are free to come back to the tasks and submit answers multiple times. In this study, we will be looking at two main variables: the total score and the total time on task in the challenge. The analysis combines these primary variables with a number of study population demographic variables: country, language, age group and gender.

The data was inspected and cleaned prior to commencing the statistical procedures. In the cleaning process, we removed test accounts and unreliable observations such as

- observations with the time on task in total being only a few seconds,
- observations with the time on task in total being over three hours, and
- observations with the starting date and time in one task being the exact same for multiple submissions (this should not be possible).

Then, the analysis continued with calculating the total scores and total times on task. After that, the quantitative statistical analysis was made and the results were analyzed.

In addition to analyzing the data from students, to collect information about the experiences of teachers a questionnaire was prepared. The questionnaire focuses on the difficulties and speed of using the challenge system with one question about the motivation of the teachers: why they participate with their students in the challenge.

3 Organizing the Challenge

In the BeLLE consortium, the aim is to gather together countries that organise the Bebras challenge to offer a free and easy-to-use platform for the challenge and to conduct multinational research on the challenge. The organiser of the challenge in each partner country is responsible for their challenge, and in 2022, Cuba, Dominican Republic, Hungary, Spain and Türkiye organised their challenge in ViLLE and Lithuania in their own system. However, the consortium also included Finland (being the creator of ViLLE), Ireland, and Sweden with intentions of organising the Bebras challenge as part of the BeLLE consortium.

This study focuses on the six BeLLE consortiun countries that collected data through the official 2022 Bebras challenge. In **Cuba**, the official challenge was organized with a plan to get all students to participate regardless of their school or study program. In **Dominican Republic**, the official challenge was organized in two languages: Spanish and English to make it possible for international students to participate. In **Hungary**, the official challenge was organized by communicating mostly with the informatics teachers who organized the challenge for all students in the school. In addition to schools in Hungary, some small schools in neighboring countries participated in the Hungarian challenge. In **Lithuania**, the Bebras challenge is quite famous and many teachers and students are eager to participate each year. In **Spain**, organizing the official challenge is difficult due to differences in different autonomous communities as well as the large number of schools that may be public, subsidized or private. Also, because of the cultural diversity in Spain, the challenge is offered in four languages: Basque, Spanish, Catalan and English. In **Türkiye**, the organized challenge was more like a pilot than an official challenge since the system was new and the process of handing out the activation codes to teachers was a big of task in Türkiye. Therefore, the number of students was not as large as for the other partners.

3.1 Preparing and Taking the Challenge

The BeLLE consortium selects a sub-set of tasks for the Bebras Challenge from those prepared annually in the Bebras community by teachers and experts in the field. The BeLLE task sub-set is then refined collaboratively by the partners and tasks that are suitable for conversion to an interactive form are implemented as interactive versions in ViLLE.

Each partner country then translates the tasks into the required languages maintaining the nature of the task as closely as possible to the English version. In 2022 the tasks were implemented in three interlinked systems, Bebras Lodge, ViLLE and the Lithuanian challenge system, since each system offers slightly different functionality. Bebras Lodge tasks are imported into ViLLE and the Lithuanian challenge system. Finally, partner countries can import the translations into ViLLE and the Lithuanian challenge system, and the challenges for each age group in each language, are constructed.

Each country determines a time frame during which the challenge is open for teachers and students. Teachers access ViLLE with an activation code: they receive information about the challenge and get an opportunity to create an account for themselves and the students. Each teacher may organise the challenge for multiple groups of students in different age groups.

Students log into ViLLE with the accounts teachers provide to engage in the challenge. Students have a time limit to complete all tasks, which is, for all partners in BeLLE, 45 min. The time limit can be changed by the teacher in case of power outages, for students with reading difficulties or other reasons.

Within the time limit, students may submit tasks as many times as needed: the latest submission is taken into account when calculating the individual scores, while all submissions are taken into account when calculating the time on task. Pupils are free to choose which tasks to do and in which order. The highest scores from the national challenges are published anonymously in ViLLE to inform teachers after each national challenge is complete.

3.2 The Structure of the Challenge

The Bebras challenge is organized in six age groups: I Pre-Primary (grades 1–2, 5–8 years old), II Primary (grades 3–4, 8–10 years old), III Benjamins (grades 5–6, 10–12 years old), IV Cadets (grades 7–8, 12–14 years old), V Juniors (grades

9–10, 14–16 years old), and VI Seniors (grades 11–12(13), 16–19 years old). Each country may decide to host the challenge only to some of the age groups. The selection of tasks in 2022 is described in Table 1.

For each age group in each country, the tasks are divided into three difficulty levels: easy, medium and difficult, and solving each task correctly gives three, nine or twelve points respectively, and incorrectly reduces points by two, three or four respectively. To make sure no student gets a negative final score, a starting score is introduced. Thus, if a student answers all tasks incorrectly, they get a final score of zero. These difficulty levels are decided by the Bebras community and reviewed by the BeLLE consortium.

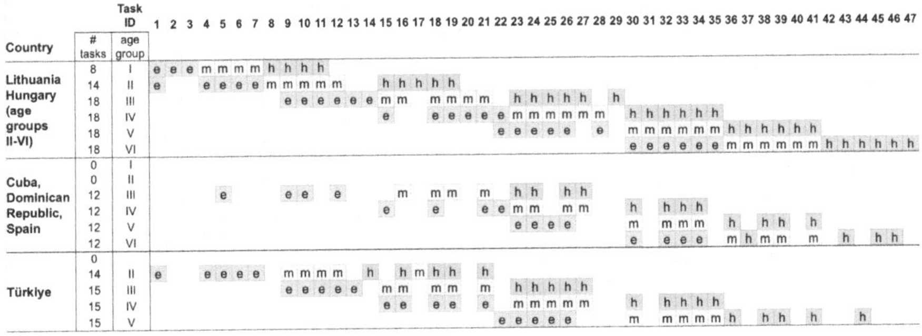

Fig. 1. Correspondence of task selection by country group("e" means an easy task, "m" an intermediate task and "h" a hard task)

4 Results

There were 94,948 observations in the six countries and a more detailed breakdown of these numbers is presented in Table 1. In Cuba, there were five provinces that failed to enroll students in the challenge which means a lower number of students participated than was expected.

Table 1. Number of students

	I	II	III	IV	V	VI	Total
Cuba	-	-	65	13	48	54	**180**
Dominican Republic	-	-	418	368	248	235	**1269**
Hungary	-	2771	7573	8006	13267	5680	**37297**
Lithuania	2376	4350	16444	12593	13018	4418	**53199**
Spain	-	-	300	777	687	280	**2044**
Türkiye	-	101	218	153	487	-	**2954**
Total	**2376**	**7222**	**25018**	**21910**	**27755**	**10667**	**94948**

Students' scores were scaled to lie in the range 0 to 1 (or 0 to 100%). The scores are not directly comparable between age groups and countries since some countries included more tasks in the sets of tasks for some age groups (as has already been discussed and summarised in Fig. 1). However, this scale gives us an overall view of students success in different age groups and countries.

Figure 2 shows the scores in all age groups in all countries, and Fig. 3 offers the same information by emphasizing the age groups. Most students have received around only 25 to 35% of the maximum scores with most students on the lower scoring side of the mean. Students from Cuba has scored slightly higher than students from Dominican Republic in age groups IV–VI, but lower in age group III. Spain has used the same task set and the students' scores are more varied and little bit higher than in Cuba and Dominican Republic. Hungary and Lithuania used the same task set in age groups III-VI (18 tasks), and Hungarian students have scored higher in all age groups.

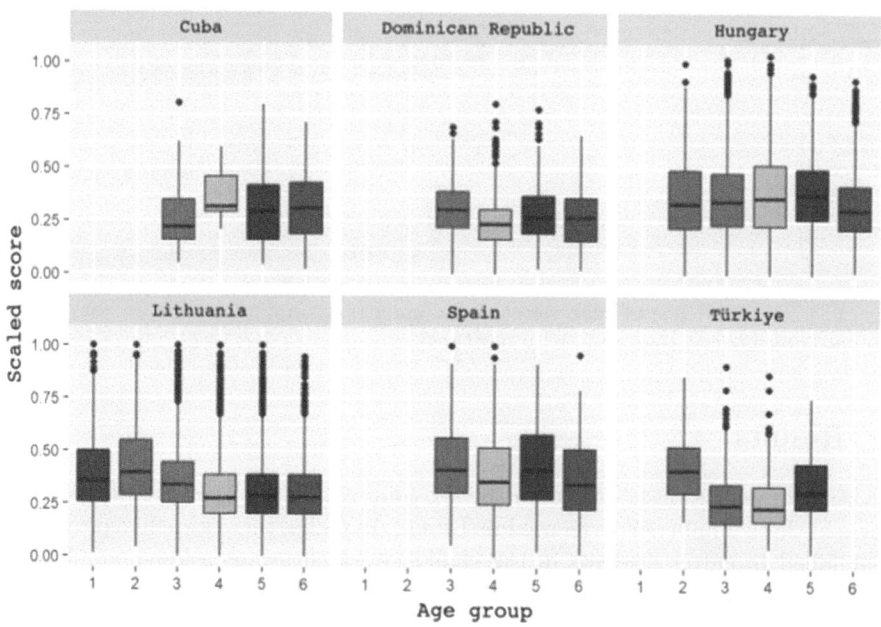

Fig. 2. Students scores by country and age group

The middle 50% of students in all age groups and countries have received low scores. Notably, only in Spain did the top 25% of students in all age groups achieve a result of over 50% of the maximum score. Spain is one of three countries to have 12 tasks for all age groups. The other two, Cuba and Dominican Republic, had generally lower means and smaller variance.

In Hungary and Lithuania, the age groups III–VI had 18 tasks, and only Hungarian age groups IV and V have the top 25% of students getting over 50% of the maximum score. In those age groups, both countries used the same set of tasks.

In age group II Hungary, Lithuania and Türkiye used the same core of 12 tasks, and Lithuania added three additional tasks. Still, the top 25% of students

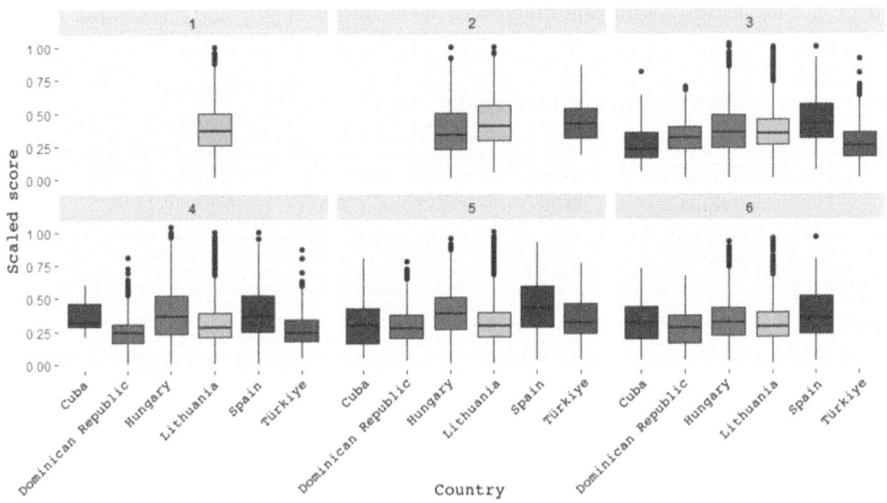

Fig. 3. Students scores by age group and country

in age group II in Lithuania and in Türkiye got clearly over 50% of the maximum score, whereas in Hungary in the same age group, there were students in the top 25% who got nearly 50% of the maximum score.

In Lithuania, also the top 25% of students in age group I got over 50% of the maximum score. In all other seventeen (excluding Spain) age groups, not even all of the top 25% of students got 50% of the maximum score.

4.1 Comparisons Between Girls and Boys

In the 2022 sample 47% of participants in all countries were girls and 53% were boys. To examine the differences between girls and boys, we performed t-tests and Mann-Whitney's U-tests to the means of the scaled scores overall and separately for each age group in each country. There was a statistically significant difference ($p < 0.001$, $n = 94904$) when we compared girls and boys for the sample as a whole. However, since the sample sizes varied substantially between countries and age groups, it is necessary to also examine groups of interest separately. Table 2 lists the statistically significant ($p < 0.001$) groups of participants.

There are statistically significant differences between girls and boys in Hungary and Lithuania only. In Hungary, there is a difference overall ($p < 0.001$, $N = 37,297$) and in groups V ($p < 0.001$, $N = 13,267$) and VI ($p < 0.001$, $N = 5,680$). In Lithuania, the difference was not showing overall, but only in groups III ($p < 0.001$, $N = 16,437$), V ($p < 0.001$, $N = 12,990$) and VI ($p < 0.001$, $N = 4,415$). The differences are small, but in Lithuania they seem to get bigger when students get older which is the opposite from Hungary (in the groups V and VI, the tasks in Hungary and Lithuania are the same).

Table 2. Statistically significant differences in the means of girls and boys by age group and country

Country	Age group	Mean (girls)	Mean (boys)	Difference	p-value	N
Hungary	all	0.368	0.383	−0.015	<0.001	37,297
Hungary	5	0.374	0.414	−0.04	<0.001	13,267
Hungary	6	0.305	0.328	−0.023	<0.001	5,680
Lithuania	3	0.368	0.353	0.015	<0.001	16,437
Lithuania	5	0.3	0.32	−0.02	<0.001	12,990
Lithuania	6	0.282	0.319	−0.037	<0.001	4,415

We also performed Chi-squared tests on the scaled scores for girls and boys in different countries (see the residuals in Appendix A). The results are aligned with the results from the t-tests and Mann-Whitney's U-tests: differences occur only in Hungary and Lithuania where boys score higher than girls.

In addition to scores, we looked at "time on task". The students were given 45 min to complete the challenge, but teachers might have given them more time. Therefore we formed a categorical variable from the times on task and divided students into ten categories: nine same-sized categories within 45 min and one category for students whose time on tasks was over 45 min. 41,7% of students used 30–40 min. (see Table 3).

Table 3. Numbers of girls and boys in each time on tasks category in all countries as a whole

Time category	Girls	Boys	Total	Total-%
0–5	224	379	603	0,6%
5–10	575	1225	1800	1,9%
10–15	1395	2478	3873	4,1%
15–20	2998	4332	7330	7,7%
20–25	5206	5973	11179	11,8%
25–30	7825	8136	15961	16,8%
30–35	10060	10184	20244	21,3%
35–40	9679	9695	19374	20,4%
40–45	5896	7024	12920	13,6%
45-...	736	928	1664	1,8%

Chi-squared tests between "time on task" to see the differences between girls and boys in different countries (see Appendix B). The results are divided into two groups. In Cuba, Dominican Republic, Türkiye and Spain, results for both girls and boys are similar. There are no significant differences between girls and boys for time on tasks categories. In Hungary, boys are generally using either less or more time than girls; meaning that there are more girls using 25–40 minutes, but more boys using less or more than that. The differences are greatest in time on tasks categories 5–20 (more boys), 35–40 (more girls) and 40–45 (more boys). In Lithuania, the trend is similar: the differences are greatest in "time on task" categories 0–20 (more boys) and 30–40 (more girls). Contrary to Hungary, the time on "task" category 40–45 is quite even in Lithuania.

4.2 Teacher Questionnaire

A questionnaire was distributed to teachers in Hungary, Spain and Türkiye. Table 4 shows the numbers of answers to the questionnaire, number of different schools that the teachers were from, the number of active teachers in ViLLE during the challenge and the percent of teachers answering the questionnaire. In Hungary, most teachers managed 2–8 groups (average 6), and in Spain 2 groups were typical (average 1.8).

Table 4. Numbers of answers to the teacher questionnaire in Hungary, Spain and Türkiye

Country	Answers	Schools	Teachers in ViLLE	Answers-%
Hungary	48	48	445	10,8%
Spain	17	14	53	32,1%
Türkiye	11	11	35	31,4%

The teachers reported in the questionnaire that students had some problems moving between the tasks and understanding if their answer was saved or not, but the teachers were mostly able to help them. In Hungary, older age groups were able to complete the challenge without any problems since they were familiar with ViLLE from the previous year. Teachers reported that ViLLE was fast and the registration process for teachers was easy to use: 40–50% in Hungary and Türkiye and 70% in Spain chose option 5 and 35–40% chose option 4 in scale 1–5. Only one teacher per country chose option 1–3.

In Hungary, the one problem mentioned was that students' account information for downloading later was missing. However, to ensure the safety of the information, students' passwords are not saved in a readable way and therefore cannot be downloaded again. In Türkiye, an additional problem was that students in a class needed to be treated together, meaning that they couldn't publish scores for the group unless all students from the group had completed the challenge. Some teachers suggested that ViLLE would be open for longer so that students could practice and get the solutions and explanations to tasks for longer.

Regarding teachers motivation for participating in the challenge; in Hungary, teachers reported that the main reasons for participation are that they like the tasks (70,8%), they think that the challenge helps to show their students CS topics (64,6%), they think that the challenge helps motivate their students (64,4%), and they think that the challenge helps getting the students' interest in CS (47,9%). Some (2%) also mentioned (with their own words) that they use the challenge as feedback on students' skills and competencies. In Spain, the teachers reported that the main incentive to participate is that it helps them to motivate their students (71%). Other reasons were that they like the tasks (41%) and they need to participate in as many competitions as they can (41%).

5 Discussion

Our first research question concerned absolute performance of the students, and we have compared the results, where the tasks performed by the students in each age group were sufficiently similar, and the time allowed to complete the challenge was 45 min or less. Lithuania and Hungary share the same task sets, and we conclude that Hungarian students scored higher than Lithuanian students in all but one (II) age group. In all challenges students typically gain between 25 and 35% of the maximum score, with the bulk of the population grouped below the mean. Thus we can conclude that the tasks are indeed a "challenge" for most participants, and that the spread of results means that many of the task items are good discriminators in terms of aspects of computing skills or perhaps evidence of ability to discern varying levels of CT ability. This hypothesis needs to be confirmed by a more thorough analysis of the tasks to categorise their focus on CT and CS skills and habits of thinking.

With the second research question, we wanted to explore the differences between girls and boys. There were some statistically significant differences in means between boys and girls in Hungary and Lithuania, but in all other countries no difference was significant. In age groups V and VI in Hungary and Lithuania, boys were scoring higher than girls, but interestingly the difference was bigger in age group V in Hungary and in age group VI in Lithuania.

In addition to the scores, we also analysed the time students have spent on task. In general, most students spent 30–40 min. However, the more interesting finding was that in Hungary and Lithuania, boys were using either less or more time than girls. It is quite difficult to know why this happens, and even though girls are using consistently more time than boys, they are also scoring slightly lower. In future work we intend to observe classes in these countries and also conduct focus-group interviews with teachers in order to generate data that may help us to understand this phenomenon better.

6 Conclusion

For historical reasons the Bebras challenge is more established in some countries and less in others, and additionally, teachers ultimately decide whether their students participate or not. It is worth noting that the challenge originated in Lithuania and has become an institution there, which explains the high participation levels across all age groups. This affects on the numbers of students who participate, and leads to vastly different numbers of students in different countries which, in turn, affects the reliability of the data. However, the results of the challenge still gives us information about the challenge and pointers to the future: what kind of differences we might expect in the future and whether the results change over time.

In this study, only one annual challenge results were analysed to answer two research questions: overall performance and gender differences. The results show that, in general, the challenge seems difficult for all students in all age groups, and several theories can be advanced to explain this outcome.

Students have a limited time to complete all tasks - three minutes per task on average - and that includes the time that they spend reading the task body. The Bebras community and the BeLLE consortium both try to limit the lengths of the tasks, but especially for younger students, even a little bit of reading might be a lot. For the older age groups, the task bodies are typically longer since the tasks get more complicated the higher the age group. All students had the same time limit (45 min), but 84,6 % of students completed the challenge in less than 40 min, which is typical especially in countries where the lessons are 45 min, because some of the lesson is always spent on guiding students.

The tasks have varied levels of difficulty that is determined by the Bebras community, the BeLLE consortium and the national organisers. Students have the opportunity to choose tasks based on the level, and that affects the overall score. Although, the intention behind the Bebras challenge is sometimes to seek out the best students, and therefore the challenge needs to be generally difficult.

Countries have differences in curricula regarding CS and CT topics, which naturally leads to differences in students' skills. For example, some countries have a separate subject of informatics, and others have included topics of informatics in other subjects such as science and mathematics [12]. In addition, topics of informatics in curricula in some countries have been added in the past decade or so, and some teachers might not feel qualified to teach them since they have not studied them themselves. The tasks also usually require a certain degree of abstraction which is difficult for students who are not sufficient in abstractions.

As Tomcsányi and Vaníček [10] have also reported, there are differences in educational culture, the text of the tasks - especially when the tasks are translated - and the age of the students. Bebras challenge consists of age groups and each of them covers two grade levels, and therefore there is variation in students' skills even within an age group due to students' ages.

7 Future Work

The Bebras challenge is an annual challenge and therefore similar studies will be made in the future as well. However, there are a couple of important improvements to be made. Firstly, students should have as similar numbers of tasks as possible, and secondly, those tasks should be the same with the same difficulty levels. That would give better insights on differences between countries. The BeLLE consortium continues to attract more countries to participate in the research consortium to enable multinational research and free and easy-to-use platform for the challenge.

Acknowledgements. This research was partially supported through the ERASMUS+ Programme – KA220-SCH - Cooperation partnerships in school education, Project Reference: 2022-1-LT01-KA220-SCH-000088736. Information about the project can be found on the CT&MathABLE website at https://www.fsf.vu.lt/en/ct-math-able.

Appendix A - Residuals of the Chi-squared tests performed to score categories against gender

Table 5.

Table 5. Residuals of the Chi-squared tests (score category and gender) in Cuba, Hungary, Dominican Republic, Lithuania, Türkiye and Spain

CU	Female	Male	HU	Female	Male	DO	Female	Male
0-0.05	0.88	−0.66	0-0.05	−0.60	0.56	0-0.05	1.06	−1.06
0.06-0.10	1.52	−1.14	0.06-0.10	−0.02	0.02	0.06-0.10	0.73	−0.73
0.11-0.15	1.20	−0.90	0.11-0.15	0.41	−0.38	0.11-0.15	0.78	−0.79
0.16-0.20	0.66	−0.49	0.16-0.20	0.08	−0.08	0.16-0.20	1.03	−1.03
0.21-0.25	−0.34	0.25	0.21-0.25	1.60	−1.50	0.21-0.25	0.13	−0.13
0.26-0.30	−0.71	0.53	0.26-0.30	1.36	−1.27	0.26-0.30	−0.75	0.75
0.31-0.35	−0.74	0.56	0.31-0.35	1.38	−1.29	0.31-0.35	0.12	−0.12
0.36-0.40	−0.59	0.44	0.36-0.40	3.04	−2.84	0.36-0.40	−0.41	0.42
0.41-0.45	−0.18	0.13	0.41-0.45	1.13	−1.05	0.41-0.45	−1.63	1.64
0.46-0.50	−0.99	0.74	0.46-0.50	−0.80	0.75	0.46-0.50	−1.18	1.19
0.51-0.55	0.46	−0.35	0.51-0.55	0.06	−0.05	0.51-0.55	0.30	−0.30
0.56-0.60	0.14	−0.11	0.56-0.60	−2.10	1.96	0.56-0.60	0.51	−0.51
0.61-0.65	−0.37	0.28	0.61-0.65	−1.25	1.17	0.61-0.65	0.66	−0.66
0.66-0.70	−0.08	0.06	0.66-0.70	−2.28	2.13	0.66-0.70	−1.38	1.39
0.71-0.75	−0.60	0.45	0.71-0.75	−2.79	2.61	0.71-0.75	−0.01	0.01
0.76-0.80	0.33	−0.25	0.76-0.80	−3.17	2.96	0.76-0.80	−0.71	0.71
0.81-0.85	−0.85	0.64	0.81-0.85	−2.91	2.72	0.81-0.85	−0.71	0.71
0.86-0.90			0.86-0.90	−3.11	2.91	0.86-0.90		
0.91-0.95			0.91-0.95	−1.91	1.78	0.91-0.95		
0.96-1			0.96-1	−1.86	1.74	0.96-1		

LT	Female	Male	TR	Female	Male	ES	Female	Male
0-0.05	−0.57	0.54	0-0.05	−0.74	0.73	0-0.05	−0.30	0.26
0.06-0.10	−0.27	0.25	0.06-0.10	−1.05	1.04	0.06-0.10	0.43	−0.38
0.11-0.15	−1.09	1.03	0.11-0.15	0.57	−0.56	0.11-0.15	0.61	−0.54
0.16-0.20	−0.93	0.88	0.16-0.20	−1.01	1.00	0.16-0.20	0.55	−0.49
0.21-0.25	−0.77	0.73	0.21-0.25	−0.44	0.44	0.21-0.25	0.01	−0.01
0.26-0.30	1.04	−0.98	0.26-0.30	1.25	−1.24	0.26-0.30	−1.08	0.97
0.31-0.35	2.09	−1.98	0.31-0.35	−0.21	0.21	0.31-0.35	−0.03	0.03
0.36-0.40	0.89	−0.84	0.36-0.40	0.12	−0.12	0.36-0.40	−0.30	0.26
0.41-0.45	2.22	−2.10	0.41-0.45	1.24	−1.23	0.41-0.45	0.18	−0.16
0.46-0.50	1.40	−1.33	0.46-0.50	0.42	−0.42	0.46-0.50	1.58	−1.41
0.51-0.55	0.26	−0.24	0.51-0.55	0.97	−0.96	0.51-0.55	0.45	−0.40
0.56-0.60	−1.49	1.41	0.56-0.60	−0.43	0.43	0.56-0.60	−0.07	0.06
0.61-0.65	−1.52	1.44	0.61-0.65	−0.86	0.86	0.61-0.65	−0.82	0.74
0.66-0.70	−1.20	1.14	0.66-0.70	−0.31	0.31	0.66-0.70	−0.66	0.59
0.71-0.75	−1.40	1.33	0.71-0.75	−0.31	0.30	0.71-0.75	0.74	−0.66
0.76-0.80	−2.28	2.16	0.76-0.80	−0.57	0.56	0.76-0.80	−2.01	1.79
0.81-0.85	−2.53	2.39	0.81-0.85			0.81-0.85	−0.25	0.22
0.86-0.90	−2.86	2.71	0.86-0.90	−1.22	1.21	0.86-0.90	−1.00	0.89
0.91-0.95	−2.13	2.02	0.91-0.95	−0.70	0.70	0.91-0.95	−0.89	0.79
0.96-1	−1.98	1.88	0.96-1			0.96-1	1.64	−1.47

Appendix B - Residuals of the Chi-squared tests performed to time on tasks categories against gender

Table 6.

Table 6. Residuals of the Chi-squared tests (time on tasks categories and gender) in Cuba, Hungary, Dominican Republic, Lithuania, Türkiye and Spain

CU	Female	Male	HU	Female	Male	DO	Female	Male
0-5	−0.11	0.09	0-5	0.36	−0.34	0-5	−0.42	0.42
5-10	0.30	−0.22	5-10	−5.45	5.09	5-10	−0.87	0.87
10-15	−0.33	0.25	10-15	−4.79	4.48	10-15	−0.96	0.96
15-20	1.08	−0.82	15-20	−3.89	3.64	15-20	−0.17	0.18
20-25	1.83	−1.38	20-25	0.73	−0.68	20-25	−0.38	0.38
25-30	−0.88	0.66	25-30	2.98	−2.78	25-30	1.98	−1.99
30-35	−0.74	0.56	30-35	2.18	−2.04	30-35	0.55	−0.55
35-40	−0.06	0.04	35-40	3.59	−3.36	35-40	0.10	−0.10
40-45	−0.56	0.42	40-45	−4.42	4.13	40-45	−0.76	0.76
45-...	−0.99	0.74	45-...	−1.59	1.48	45-...	−1.23	1.24

LT	Female	Male	TR	Female	Male	ES	Female	Male
0-5	−5.01	4.74	0-5	0.64	−0.63	0-5	−1.41	1.26
5-10	−7.64	7.23	5-10	−0.30	0.30	5-10	−0.98	0.87
10-15	−9.06	8.57	10-15	0.25	−0.25	10-15	−0.79	0.71
15-20	−6.84	6.47	15-20	−0.76	0.75	15-20	−0.06	0.06
20-25	−1.41	1.33	20-25	−0.28	0.27	20-25	−0.07	0.06
25-30	2.31	−2.19	25-30	0.46	−0.46	25-30	−0.44	0.39
30-35	5.73	−5.42	30-35	0.36	−0.36	30-35	0.44	−0.40
35-40	5.20	−4.92	35-40	−0.92	0.91	35-40	0.62	−0.55
40-45	0.57	−0.54	40-45	0.58	−0.58	40-45	0.89	−0.79
45-...	−1.08	1.02	45-...	1.24	−1.23	45-...	−0.18	0.16

References

1. Bagger, A., Vennberg, H.: Assessment of mathematics in Preschool-class. In: NORSMA10 Paper Session 5: Equal Access for All Learners to Quality Mathematics Education (2021)
2. Bilbao, J., et al.: Algebraic thinking and computational thinking in pre-university curriculum. In: INTED2023 Proceedings, pp. 3888–3895 (2023). https://doi.org/10.21125/inted.2023.1037

3. Dagiene, V., Stupuriene, G.: Bebras - a sustainable community building model for the concept based learning of informatics and computational thinking. Inform. Educ. **15**(1), 25–44 (2016)
4. Dagienė, V., Stupurienė, G., Vinikienė, L.: Implementation of dynamic tasks on informatics and computational thinking. Baltic J. Mod. Comput. **5**(3), 306–316 (2017)
5. Kurvinen, E., Dagienė, V., Laakso, M.-J.: The impact and effectiveness of technology enhanced mathematics learning. In: Dagienė,V., Jasutė, E. (eds.) Constructionism 2018: Constructionism, Computational Thinking and Educational Innovation: Conference Proceedings. Vilnius University, pp. 351–363 (2018)
6. Kurvinen, E., Kaila, E., Laakso, M., Salakoski, T.: Long term effects on technology enhanced learning: the use of weekly digital lessons in mathematics. Inf. Educ. **19**(1), 51–75 (2020)
7. Laakso, M.-J., Kaila, E., Rajala, T.: ViLLE - collaborative education tool: designing and utilizing an exercise-based learning environment. Educ. Inf. Technol. **23**(4), 1655–1676 (2018)
8. Niemelä, P., Pears, A., Dagienė, V., Laanpere, M.: Computational thinking – forces shaping curriculum and policy in Finland, Sweden and the Baltic countries. In: Passey, D., Leahy, D., Williams, L., Holvikivi, J., Ruohonen, M. (eds.) Digital Transformation of Education and Learning - Past, Present and Future: IFIP TC 3 Open Conference on Computers in Education, OCCE 2021, Tampere, Finland, August 17–20, 2021, Proceedings, pp. 131–143. Springer International Publishing, Cham (2022). https://doi.org/10.1007/978-3-030-97986-7_11
9. Pluhár, Z., Kaarto, H., Parviainen, M., Garcha, S., Shah, V., Dagienė, V., Laakso, M.-J.: Bebras challenge in a learning analytics enriched environment: Hungarian and Indian cases. In: Bollin, A., Futschek, G. (eds.) Informatics in Schools. A Step Beyond Digital Education: 15th International Conference on Informatics in Schools: Situation, Evolution, and Perspectives, ISSEP 2022, Vienna, Austria, September 26–28, 2022, Proceedings, pp. 40–53. Springer International Publishing, Cham (2022). https://doi.org/10.1007/978-3-031-15851-3_4
10. Tomcsányi, P., Vaníček, J.: International comparison of problems from an informatic contest. In: ICTE 2009 : Information and Communication Technology in Education 2009. - Ostrava : University of Ostrava, 2009. - ISBN 978-80-7368-459-4. - S. 219–221
11. UNESCO article. https://www.unesco.org/en/articles/unesco-prize-awarded-collaborative-learning-platform-ville-finland (6 April 2021). Accessed 9 June 2023
12. Yeni, S., Grgurina, N., Saeli, M., Hermans, F., Tolboom, J., Barendsen, E.: Interdisciplinary integration of computational thinking in K-12 education: a systematic review. Inform. Educ. **23**(1), 223–278 (2024). https://doi.org/10.15388/infedu.2024.08

Innovative Teaching Beyond the Classroom

From Caesar Shifts to Kid-Enigma. The CS Unplugged-Like Path in the MuMa Science Centre

Michał Ren[1](✉)[iD], Paweł Perekietka[2], and Łukasz Nitschke[3]

[1] Adam Mickiewicz University, Wieniawskiego 1, 61-712 Poznań, Poland
renmich@amu.edu.pl
[2] Fundacja Zakłady Kórnickie, Flensa 2b, 62–035 Kórnik, Poland
[3] Kórnik, Poland

Abstract. We present a proposal for a learning path for children to teach them cryptography during visits to the MuMa Science Centre. As the goal of the Math Museum (MuMa) for which the track is being developed is to make youngsters enthusiastic about mathematics and computer science, we have decided to focus on cryptography. In our past experience this proved to be very effective – children love secrets and spy games. We will prepare a set of CS-unplugged-like activities which will cover the broad range from the simplest, historical ciphers up to the Enigma cipher. In contrast to previously described CS-unplugged activities, we will focus on methods of breaking those ciphers, not on merely using them. Breaking of Enigma deserves attention thanks both to clever use of pure mathematics, but also due to its historical significance in ending WWII. In this paper we will present ideas for activities to teach cryptanalysis of the simplest ciphers, starting from Caesar and Vigenère ciphers, as well as the design for a simple, paper teaching aid that simulates the simplified Enigma to show its properties. We share pertinent feedback we have received after several presentations we had already made to small groups of children and adults.

Keywords: CS Unplugged · Enigma · cipher · cryptanalysis · cryptography · science centre

1 Introduction and Related Works

Activities related to cryptology are an important part of the Computer Science Unplugged project, in which students from elementary school ages upwards work without computers with hands-on activities that help them understand a broad range of topics in an engaging and motivating way. This approach is a good fit for institutions that provide informal education outside of classrooms, such as science centres, like the one being built near Poznań, Poland. One of the areas it will focus on will be cryptology, especially the breaking of the Enigma cipher before WWII by alumni of Adam Mickiewicz University – M. Rejewski,

J. Różycki and H. Zygalski. The role of Polish cryptologists is often overlooked, but some recent publications have explored the subject more fully [1,2].

We are going to design a learning path around breaking of the Enigma cipher. In this paper, we share the results of our conceptual work and summarize pertinent feedback that we received after trials of some of the planned activities.

Mike Fellows and Neal Koblitz [3] described how notions of cryptography using combinatorial constructions had been successfully presented to youngsters in the late 1980 s. Their goal was to provide topics of modern cryptography as "a stimulating context for logical and mathematical modes of thinking" [4]. Importantly, no computers were used. At that time the term Kid Krypto was also coined, referring to "the development of cryptographic ideas that are accessible and appealing (and moderately secure) to those who do not have university-level mathematical training" [5].

There are many publications that present popular ciphers, but few that show how to break them at a level suitable for children. While Koblitz has proposed to cryptanalyze an RSA-like cipher ("Kid-RSA") [4] we would like to present the Enigma cipher in an accessible way, as its breaking had huge influence on WWII, impacting millions of people. Unfortunately, the most accessible complete description of breaking the Enigma known to us is by Rejewski himself [7], and it requires knowledge of permutation theory, as well as mathematics at least at college level. It is an interesting question how deeply we can explain to children how Enigma worked and how it was broken without trivializing the topic. We aim to achieve that with simplified examples.

2 From Caesar to Enigma

In the following subsections, we will describe the activities that will form the crucial parts of the cryptology learning path in the MuMa Science Centre.

2.1 Caesar

The first introductory activity is solving of a classic Caesar cipher, which is a substitution cipher created by shifting every letter three places forward in the alphabet. To facilitate work in groups and to prepare for further activities, a visual aid is printed on the handouts, consisting of the alphabet printed twice: ABCDEFGHIJKLMNOPQRSTUVWXYZABCDEFGHIJKLMNOPQRSTUVWXYZ at the very top and bottom of the page, so that two handouts can be put together to produce any required shift. Depending on age and sophistication of the group, that technique can be demonstrated, or students can come up with the idea on their own. It is usually sufficient to provide a single encrypted sentence to be decrypted as a challenge.

The second activity is to decrypt a longer message, enciphered with a Caesar cipher with an unknown shift. The encrypted message should preserve spaces between words and punctuation. This encourages students to seek ways to make finding the key (the shift) easier (such as focusing on single-letter words, etc.), which is an important concept for more difficult ciphers. We have found the desire to avoid drudgery to be a powerful inspiration for creative thinking.

2.2 Vigenère

The Vigenère cipher can be thought of as a series of Caesar ciphers with different shifts. The key is a series of numbers; e.g. with key (1,2,3), the first letter will be shifted by 1, the 2nd by 2, the 3rd by 3, the 4th by 1, the 5th by 2, etc. (repeating)

A longer encrypted message is provided, with known key length. It is difficult to decrypt even when punctuation and spaces between words are left in. One technique that makes it trivial is frequency analysis – high school students are able to discover it on their own, but for younger groups it should be explained, as it is vital for more difficult ciphers. One can leverage different letter frequencies in natural language – if the key is of length n, by trying to shift every n^{th} letter (all sharing the same shift) by different values, and looking when the resulting letters fit the frequency distribution of the language of the message.

2.3 Monoalphabetic Substitution Cipher

To make sure that students are able to use statistical ideas to solve ciphers, they are given a monoalphabetic substitution cipher challenge, in which each letter is replaced with another (but always the same one). This challenge should not preserve spaces between words, nor punctuation. There are $26! \approx 10^{26}$ possible keys (which high-schoolers can calculate), so using brute force is not possible. Various "crossword techniques" (trying to guess words, looking for double letters, etc.) help little, due to lack of punctuation. However, if the message is of sufficient length (one or two paragraphs), solving it with frequency analysis is easy.

2.4 Kid-Enigma

In this activity, a device made of paper, called a stepping reflector, is used to generate an Enigma-like cipher. It consists of two wheels – a larger, stationary wheel with 26 letters of Latin alphabet on its circumference, and a smaller, spinning wheel, with lines which connect letters into pairs. There is an arrow on the smaller wheel which, at a given position, points to one of the letters, to easily refer to the current position, as "setting A", "setting B", etc.

Encryption using this toy version of Enigma consists of substituting a letter by the one connected to it by a line on the wheel, after which the small wheel is turned $1/26^{th}$ of the full angle. Decryption is done in exactly the same way!

It is crucial to note that the Enigma cipher was reciprocal: at a given setting if X would be encrypted to Y, then Y would be encrypted to X (so, no letter was ever encrypted to itself). Also, with the same setting, an Enigma machine could be used for both encryption and decryption. The cipher obtained by using the stepping reflector is reciprocal as well. This property has been omitted in some explanations of the Enigma in popular literature [8], but it allowed Rejewski and

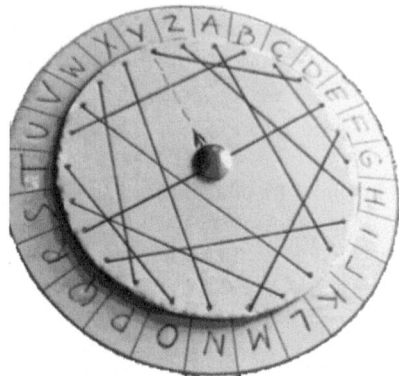

Fig. 1. The stepping reflector – a toy example of the Enigma machine

his colleagues to use the permutation theory to attack the Enigma cipher [7], so we consider its accurate presentation to students as very important.

Our toy Enigma represents the concept of stepping well – it changes the substitution used for successive letters of the message, similar to the real Enigma machine. Students usually easily draw parallels to the Vigenère cipher.

In the first challenge, students are familiarized with operation of the stepping reflector. Working in small teams, one person encrypts a word of their choice, then another decrypts it. A discussion should follow, to make sure that students understand what the key is in the Kid-Enigma cryptosystem – both the lines drawn on the wheel, and the initial setting of the wheel.

In the second challenge, students receive a ciphertext consisting of around 100 words. It has been encrypted as follows: before encrypting each word, the operator reset the rotor to setting A. (Similar to the way commercial Enigma machines were sometimes incorrectly reset for every message.) Students should be given some tips; we suggest lists of the top five first letters of most common words, and the most common two-letter words. Counting first letters of encrypted words (i.e. frequency analysis of letters) enables discovery which letters are connected in pairs on the wheel, and then – decryption of all the words. Note that besides easing cryptanalysis, a benefit of restarting encryption with every word is that they can be worked on in parallel by different students.

In the final challenge students receive a ciphertext consisting of a few dozen words. However, this time it has been encrypted as follows: before encrypting each word, the operator picked a setting on the rotor, and then (always starting from setting A) first encrypted the chosen setting twice, and only then set the wheel to the setting he picked, and encrypted the word that is the actual message. The result will therefore have two letters that correspond to the setting he picked, followed by the encrypted word. Some students might immediately have figured out that the words are too long, but the instructor should make sure everyone is aware how the cipher was created.

This is in fact how Enigma was actually used (settings were labeled by three letters, and chosen for the whole message) – the doubling of transmission of the chosen setting safeguarded against random errors, etc. [7] It is a good opportunity to mention historic facts about breaking the Enigma cipher in 1930s to students.

Students should be able to figure out that sending the setting twice creates a vulnerability and ask themselves whether the connections on the wheel can be derived from analysis of the sets of extra letters.

For the students to gain insight how to recreate the lines on the wheel, they should analyse in what way reciprocal substitutions change after a given letter is encrypted. Since there are 26 dots on the rim of the cipher wheel and the lines connecting pairs of dots are not known, the current input letter, e.g. A will be transformed to a letter indicated by one of the 25 dots. We cannot predict the output letter unless we know the line segments on the wheel, but let us suppose that it is the letter D. If the wheel now turns, the line that connected the letters A and D will move one position and will now connect the letters B and E. Likewise, if the letter B were connected to the letter G before the wheel turned then the letter C will be connected to the letter H after the wheel has turned. Such an analysis leads to noticing the pattern in relationships between the letters and eventually to the idea that we can recreate the substitutions (i.e. the lines on the wheel) for settings A and B. This is reminiscent of Rejewski's attack on the message keys of Enigma [7].

For a more detailed learning scenario, including examples, see the Appendix.

3 Evaluation

Through its inclusion in the ACM K-12 curriculum in 2003, the CS Unplugged-style approach started to be seen worldwide as the basis of a philosophy for teaching computing (including outreach) without using a computer [9].

Wide international interest and collaboration between educators led to a proposal of a design pattern useful for evaluating whether an activity fits in with the Unplugged-like approach [10]. We have checked our activities against that pattern, and have found that they conform to most of the desirable characteristics: (1) they use no computers but only inexpensive equipment, (2) they tell a story to engage students, (3) they involve interaction with other classmates (cooperation and competition), (4) they encourage students to discover answers by trial and error, (5) they are focused on outreach rather than teaching in schools [9].

Some of the presented activities (2.1–2.3) have already been extensively refined through feedback from various computing lessons and university science festivals, from students of various age groups. The Kid-Enigma activity (2.4) has so far been presented to interested professional educators, as well as during the CMSC 2022 conference; testing with students will commence soon. The universal appeal of the historical background, shared by all age groups, surprised us the most. We advise instructors to heavily underline how math used for problem-solving (computational thinking) can affect historical events.

4 Future Work

The concept of a stepping reflector can be used to design cryptology activities that can demonstrate brute force attacks or attacks that involve guessing expected words. It can also form the basis for presentation of the main idea behind cryptanalysis of the real Enigma in a much more accessible form – still however requiring college level mathematics. We hope to invent a strong attack on the device that will involve only high-school curriculum.

After fully developing the Enigma path in the MuMa Science Centre, we intend to further expand its cryptography section. Symmetric ciphers are not the only ideas from cryptography that can be easily explained, and serve to introduce mathematical concepts. We will eventually include several others; among them hopefully public-key cryptography (particularly RSA, as it uses relatively simple algebra), hashing and blockchain concepts, visual cryptography, secret sharing and zero-knowledge proofs.

Acknowledgments. We would like to thank Małgorzata Bednarska-Bzdęga for her insightful comments on the Kid-Enigma.

Appendix

Kid-Enigma learning scenario

Age group: Junior high school and above
Presumed abilities of participants: Familiarity with Caesar shift ciphers

1. The activity is framed in the context of a story about scouts.
   Dozens of teams of three have applied for a course on ciphers for scouts. Preliminary interviews are about to begin. Now it's time for Fran's team.

   ```
   VS NA RARZL VF NOYR GB QVFPBIRE GUR FRPERG BS N PVCURE
   GURA UR BE FUR VF FNVQ GB UNIR PENPXRQ BE OEBXRA VG
   ```

2. The students are given about 5 min to solve Fran's team's cryptogram. The teacher asks a volunteer to present a method for finding the shift (13). (A solution by guessing is: N and NA represent the articles A and AN.) Students are divided into teams and work to recreate the cipher alphabet:

   ```
   ABCDEFGHIJKLMNOPQRSTUVWXYZ
   NOPQRSTUVWXYZABCDEFGHIJKLM
   ```

   Finally, the teams work on their own to decipher the whole message:

   ```
   IF AN ENEMY IS ABLE TO DISCOVER THE SECRET OF A CIPHER
   THEN HE OR SHE IS SAID TO HAVE CRACKED OR BROKEN IT
   ```

3. The activity continues with the story about the course for scouts.

   The cipher alphabet used in preliminary task is written on the board.
   "Can you see anything special in this alphabet?," asks the instructor.
   "A stands for N and N for A; B for O and O for B; etc."
   *"Yes! The idea was to make the cipher **reciprocal**," states the instructor, and adds, "A Caesar with a shift of 13 we call a half-reversed alphabet".*

4. Each team is given a sheet of cardboard with two circles printed on it. The teacher explains how to make a device called a stepping reflector.

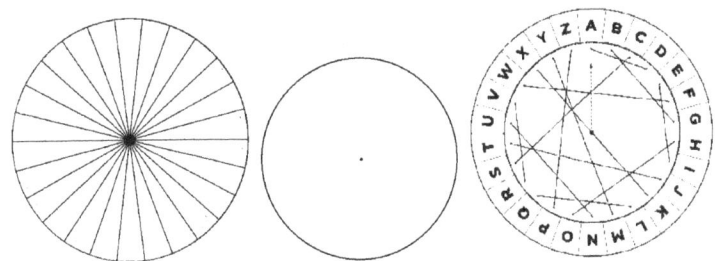

   The encryption method (algorithm) is as follows:
   (1) Turn the wheel until the arrow points to the letter A.
   (2) Look at the first letter of the word to encrypt, select the appropriate line, follow the line and note down the letter at the other end of the line.
   (3) Before encrypting the next letter, turn the wheel one step clockwise ...

5. The teams encrypt a word of their choice and decrypt the ciphertext to see how useful the reciprocal ciphers are: encryption is the same as decryption.

6. The teacher poses a question about a number of reciprocal cipher alphabets. If the students are advanced enough, it is worth encouraging them to work out the value themselves, otherwise the answer is given: $26!/(2^{13} \cdot 13!) \approx 10^{13}$.

7. The second part of the activity begins with handing out a cryptogram.

```
ZKJ JPD XZTMIWZ YBVI GK KKJPP QDWJD VY VVTXDWNSXR YS PQICPJX XZIN
KKICPJXZQ JQHGFBZ YBVI VO RSP YGZZ GO USWS GO YBJ KKICPJXDIT ZL
KPIN BSFJ XYVCHW QGLP VIJHHUHFX VKCIONW DSHTHUBH ZL YBJ BBJZG YBJ
QDFGH UQPIKOP ZL BSFJD MZHI GO XQIXQY UQPIKOP VKYZYTIIJTWRV GKY
YBCCS ZKJG GKY XS ZK GGJ KKICPJXZQ VK YBJ XPGZ BPD YBJ HDTNKO HPW
WQ WGLFKA WT G QGJUJTIPN GKVYPWNB BDPN G XZRKOXNZYNGC UPFSK RZGWKO
ZL HGDMHRMLFYF YBJ  HGDMHNIFXMFG VO GHOZ YS PDHRIKZL YBJ DFVTQEZMC
WZP GFHX YBJ DPPIKOI ZL YBJ HSWHKXXDITF WQPBKTI YBJ UQPIKOP
```

A group discussion ensues about how to attack the cryptogram. If the brainstorming stalls, the teacher says to focus only on the first letters of words. The following tips are then presented by the teacher:
   – The top letters at the beginning of English words are: T, A, I, S and O.
   – The most common two-letter words are: TO, OF, IN, IT and IS.
Together, the teams should count all the first letters of the encrypted words.

```
A B C D E F  G H I J K L M N O P Q R S T U V W  X  Y Z
0 5 0 3 0 0 12 5 0 2 5 0 1 0 0 24 2 0 0 5 8 5 7 16 8
```

The sequence of first observations about the cryptogram may be as follows:
- The most frequent letters are Y and G, followed by V, Z, and X.
- The only single letter is G.

The teacher asks: "What's the conclusion?" ("Surely G is A, and Y is T").
The competitive phase begins. The teams should no longer work together, but try on their own to recreate all lines connecting letters on the wheel. The solution (the cipher key) shall be written on the board or flipchart.
Students are then asked to decrypt the message. The teams can cooperate.

```
ONE MAY SUPPOSE THAT AN ENEMY FINDS IT IMPOSSIBLE TO DECRYPT SUCH
ENCRYPTED MESSAGE    THAT IS NOT TRUE AS LONG AS THE ENCRYPTION OF
EACH WORD STARTS FROM IDENTICAL INITIAL POSITION OF THE WHEEL    THE
FIRST LETTERS OF WORDS JUST AS SECOND LETTERS INDEPENDENTLY AND
THIRD ONES AND SO ON ARE ENCRYPTED IN THE SAME WAY    THE CIPHER CAN
BE BROKEN BY A FREQUENCY ANALYSIS WITH A SUFFICIENTLY LARGE NUMBER
OF CRYPTOGRAMS    THE CRYPTANALYST IS ABLE TO DISCOVER THE PLAINTEXT
BUT ALSO THE PATTERN OF THE CONNECTIONS BETWEEN THE LETTERS
```

8. The teacher mentions some historic facts. Here is a suggested text to tell.
   To prevent the use of statistical attacks (letter frequency analysis), i.e. exploitation of a large number (about 60–80) of cryptograms intercepted on the same day, the Enigma cryptosystem in 1930s used a three-letter message key: indicators of the initial positions of three rotors in the machine, chosen by the sender. There is some chance of an operator error or message corruption during transmission, so the three-letter key (e.g. DVR) was doubled (e.g. DVR DVR), encrypted using the daily key and sent as a message header (e.g. FSB QCH). The recipient decrypted the key (DVR) and set up the Enigma machine to read the message. Using an Enigma machine itself to encrypt the message key was a vicious circle... A gifted Polish mathematician Marian Rejewski made good use of this flaw. We will follow in his footsteps.
9. The teacher initiates a discussion on how to use the Kid-Enigma device more securely, directed in such a way as to reach the following conclusion:
   If the initial position of the wheel (used for the first letter of a word) is continually changed from word to word, the method of frequency analysis will fail.
10. The last part of the activity begins. The teachers tells the following story:
    A secret radio message sent in Morse code has been intercepted. Our intelligence supposes that Enigma-like ciphers are still being used, but a new encryption algorithm has been applied to prevent statistical attacks." "You take on roles of the cryptologists"
    The cryptogram is handed out.

```
QLDBYU PIGJJSRN WHSR MWKUAZ HXVHS NGHNZYW KQBUWR AROZ
UFLJUGSIEYSYQ JVNT XOJWQH YYBM RKDZK ZSRJLN CBYYNDR FEFWBE
DAWUYXBS VDJSVLVNMNIU ENFJIS ZSX WHGJMGD FEFJ ENHAIKZWR
```

The teacher states: "In the new algorithm the initial setting changes from word to word. We call it the word key. Before encrypting each word, the operator (starting from setting A) first encrypts the word key twice and then encrypts the word (starting from the setting indicated by the word key)."

The students should now analyse how a substitution of the Kid-Enigma changes after a letter is encrypted. It will lead them to notice the pattern in the relationships between the two letters used to encrypt the word key. The teams may start with QL, so L should be written in the grid below Q. Then they should look for AR in the cryptogram. The next one is CB, and so on. Eventually they recreate two cipher alphabets, i.e. the substitutions (the lines on the wheel) for settings A and B. Finally, each team writes a duplicated alphabet on grid paper and cuts it out. Then they try to match the cipher alphabets with the duplicate alphabet:

```
ABCDEFGHIJKLMNOPQRSTUVWXYZABCDEFGHIJKLMNOPQRSTUVWXYZ
QACVMEUJRZDFNXYLIBSTOGWHPK
LRBDWNFVKSAEGOY
```

Each team is tasked with deciphering one or two message keys and the associated word(s). The teams put together the decrypted words.

```
YOUR SUPPLY OF FOOD AND WATER WILL BE REPLENISHED IN FIVE OR SIX
DAYS   UNTIL THEN RATION EVERYTHING   SEND A REPLY TO CONFIRM
```

References

1. Welchman, G.: From Polish Bomba to British Bombe: the birth of ultra. Intell. National Secur. **1**(1), 71–110 (1986). https://doi.org/10.1080/02684528608431842
2. Grajek, M.: Enigma. Bliżej prawdy. Rebis, Poznań (2007)
3. Fellows, M., Koblitz, N.: Combinatorially Based Cryptography for Children (and Adults). Congressus Numeantium **99**, 9–41 (1994)
4. Koblitz, N.: Cryptography as a teaching tool. Cryptologia **21**(4), 317–326 (1997). https://doi.org/10.1080/0161-119791885959
5. Fellows, M., Koblitz, N.: Kid Krypto. In: Brickell, E.F. (eds.) Advances in Cryptology – CRYPTO' 92. LNCS, vol. 740. Springer, Berlin, Heidelberg. (1993). https://doi.org/10.1007/3-540-48071-4_27
6. Reeds, J.: A critical review of the book: "The Code Book: The Evolution of Secrecy from Mary, Queen of Scots to Quantum Cryptography", Notices of the American Mathematical Society, **47**(3), 369–372 (2000)
7. Rejewski, M.: Wspomnienia z mej pracy w Biurze Szyfrów Oddziału II Sztabu Głównego w latach 1930–1945 / Memories of my work at the Cipher Bureau of the General Staff Second Department 1930–1945, 2nd edn. Wydawnictwo Naukowe UAM, Poznań (2021)
8. Du Sautoy, M.: The Number Mysteries: A Mathematical Odyssey through Everyday Life. 4th Estate, London (2011), pp. 189–193 ISBN: 978-0007309863
9. Bell, T., Curzon P., Cutts, Q., Dagiene V., Haberman, B.: Overcoming obstacles to CS education by using non-programming outreach programmes. In: Proceedings of the 5th International Conference on Informatics in Schools: Situation, Evolution and Perspectives, ISSEP 2011, pp. 71–81. Springer, Berlin, Heidelberg (2011). https://doi.org/10.1007/978-3-642-24722-4_7
10. Nishida, T., Kanemune, S., Idosaka, Y., Namiki, M., Bell, T., Kuno, Y.: A CS unplugged design pattern. ACM SIGCSE Bull. **41**(1), 231–235 (2009)

Large and Parallel Human Sorting Networks

Stefan Szeider(✉)

Algorithms and Complexity Group, TU Wien, Vienna, Austria
sz@ac.tuwien.ac.at

Abstract. This paper presents two innovative extensions of the classic Human Sorting Network (HSN) activity from the CS Unplugged program. First, we describe the implementation of a large-scale HSN with 50 input nodes, realized with high school students in Vienna, Austria. We detail the logistical challenges and solutions for creating an HSN of this magnitude, including location selection, network layout, and participant coordination. Second, we report on using parallel 6-input HSNs, which introduce a competitive element and enhance engagement. This parallel setup allows for races between teams and can be adapted for various age groups and knowledge levels. Both extensions aim to increase the educational impact and enjoyment of the HSN activity. We provide comprehensive insights into our experiences, enabling other educators and researchers to replicate or further develop these HSN variants.

Keywords: Computational Thinking · Human sorting networks · CS Unplugged · algorithmic education · large-scale computing activities · parallel sorting algorithms

1 Introduction

The classic book *Computer Science Unplugged* [4] includes activities, games, and puzzles suitable for people of all ages and backgrounds to communicate fundamental concepts of computer science without the help of a computer. A quintessential activity of CS Unplugged, conceived by Mike Fellows in the late 1980s [3], is to use a *sorting network* (often with 6 input nodes) drawn on the ground, where participants hold cards with numbers (or other linearly ordered objects) and perform the sorting represented by the sorting network by stepping from node to node. At the beginning, the participants stand on the input nodes of the sorting network, and after a signal to start, they work through the network until they arrive at the output nodes holding their cards in an ordered sequence. The elementary step performed throughout the activity occurs when a pair of

---

This work was partially created while the author was a visiting scientist at the Simons Institute for the Theory of Computing. The author acknowledges the support by the Austrian Science Funds, project 10.55776/P36688 and the Vienna Science and Technology Fund (WWTF) under grant ICT19-065.

© The Author(s) 2025
H. Fernau et al. (Eds.): CMSC 2024, LNCS 15229, pp. 194–204, 2025.
https://doi.org/10.1007/978-3-031-73257-7_16

participants arrive together at a comparator node, and according to the relative ordering of the two cards they hold, the one with the smaller number continues to the next node via the *minor* edge, and one with the higher number continues to the next node via the *major* edge. Which edge is minor or major can be indicated by the color and/or thickness of the edge.

Figure 1 shows a simple example of a sorting performed on a sorting network. We refer to this activity as a *human sorting network*, or *HSN*, and to the participants as *players*.

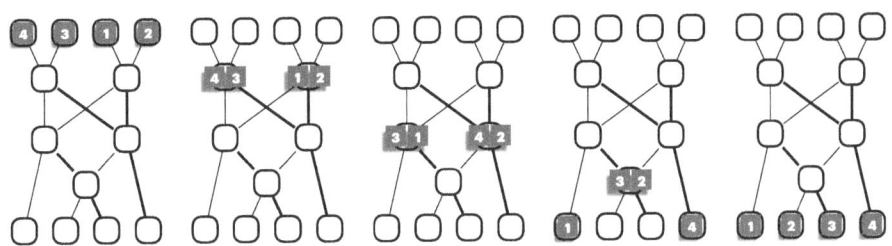

Fig. 1. Five snapshots of the sorting performed on a sorting network with 4 input nodes. Major edges are indicated by thick lines. All edges are assumed to be directed from top to bottom, also in other illustrations below.

This activity has shown to be quite successful, educational meaningful, and fun. By choosing the right type of objects to be sorted, the activity can be adjusted to the age or proficiency of the players: one-digit or 3-digit numbers, words to be sorted lexicographically, or even very domain-specific objects (see Sect. 4).

In this paper, we report on two extensions of the basic HSN activity. Our objective is to encourage educators to organize simlar events, and by laying out our experience to provide useful information.

In Sect. 3, we report on a large HSN with 50 input nodes that we realized with school classes in Vienna, explaining the challenges, technical consideration, implementation, and the overall experience. In Sect. 4, we report on the use of two small HSNs laid out next to each other, which supports the race of two groups of players and elevates engagement. In Sect. 5, we discuss extensions of these two activities that would be interesting to pursue in the future, and conclude in Sect. 6.

We hope that this report will encourage educators to organize similar events and that our experience and considerations will be useful for such endeavors.

2 Sorting Network Theory

Sorting networks are data-oblivious sorting algorithms, meaning that the sequence of comparisons performed is not influenced by the input list. This characteristic makes them ideal for implementation in hardware circuits, where the size determines the number of gates required and the depth determines the circuit's delay, but sorting networks are also used for propositional encodings of cardinality constraints [6]. Knuth [7] and Parberry [8] provide comprehensive overviews of sorting networks.

Sorting networks are often depicted as *Knuth diagrams* that show parallel lines, called *channels*, instead of input nodes, and lines connecting channels representing comparator nodes. Figure 2 shows the sorting network of Fig. 1, both as a directed acyclic graph and as a Knuth diagram.

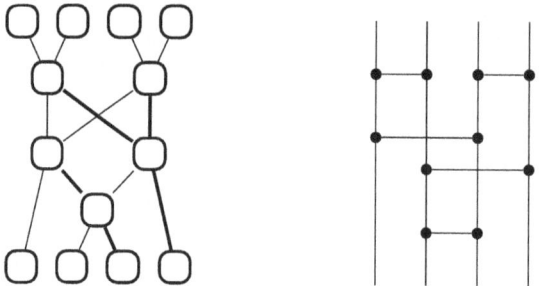

Fig. 2. Left: sorting network from Fig. 1 as a directed acyclic graph. Right: the same sorting network as a Knuth diagram.

Comparator nodes that do not depend on each other can be executed in parallel and are therefore combined into *layers*. The *depth* of a sorting network is its number of layers, which equals to the largest number of comparator nodes that lie on a path from an input node to an output node, hence the depth is also called the *delay time* of the network. For instance, the sorting network in Fig. 2 has depth 3. The *size* of a sorting network refers to its number of comparator nodes. Since the introduction of sorting networks, researchers have been seeking to construct sorting networks, aiming at minimal depth and minimal size networks. Asymptotically, there exist sorting networks of logarithmic depth and size $O(n \log(n))$ for n input nodes [1,9,10]. However, the constants hidden in the asymptotic expression are huge, so that in any practical setting, one rather resorts to other constructions, like *Batcher's bionic sorting networks* [2] of depth $O(\log^2(n))$ and size $O(n \log^2(n))$. The very simple *odd-even transposition sorting networks* [7], on the other hand, have depth n and size $n(n - 1/2)$. Optimal networks are only known for a small number of inputs[1] as one needs to exhaustively search a quickly growing search space. Interestingly, for $n \leq 11$, there exist sorting networks that are simultanously of minimial depth and minimal size, for $n \geq 12$ one needs to decide which of the two measures one wants to minimize.

---

[1] A list of optimal sorting networks for small n can be found at https://bertdobbelaere.github.io/sorting_networks.html.

3 A Giant HSN

To stimulate public awareness of algorithmic thinking and create an engaging team project for high school students, we implemented a large-scale HSN with 50 input nodes in Vienna, Austria, in September 2018. This ambitious endeavor presented several unique challenges in terms of logistics, design, and coordination. In this section, we outline the key considerations and solutions we developed to successfully realize this giant HSN. We provide sufficient detail to make this event reproducible.

Fig. 3. Photos from the giant HSN event in Vienna.

1. The *location* for setting up the HSN needs to be sufficiently large, ideally with paved ground, easily accessible by public transport, and providing facilities like lockers and washrooms for the participants. After location scouting in Vienna, we settled on the free space next to a large sports stadium.
2. We invited about eighty *high school students* from the region, mostly from the age group 14–16, to participate in the event which included helping to set up the network and performing as players.
3. For the *timing*, we planned the event to fit into one day, including set up and removal. This way, we only needed to rent the space for one day, and the participating students required only one day out of school. A day in September, just after school starts, with expected warm weather, seemed a good choice.

4. The total length of all the edges would be several miles; hence, drawing edges would be too time-consuming and challenging to clean up afterward. Hence, we decided on an *edgeless* network layout.
5. For the nodes, we used *ceramic square tiles*, size 34×34 cm, in different colors, which are placed on the ground and can easily be picked up afterwards without any residue to clean, making the setup even weatherproof.
6. We used an *odd-even transposition sorting network* (see Sect. 2) so that all the comparator nodes can be placed along a rectangular grid. Moving from one transition node to the next, the player only needs to move to an adjacent grid layer, where the larger number moves left and the smaller number moves right. Figure 4 shows such a network with 8 input nodes, once with edges drawn and once edgeless.
7. To provide additional orientation so that players do not get lost, we applied a *color coding* to the comparators nodes: the larger number stays on the same color, the smaller number doesn't (unless the left or right border of the network is reached), see Fig. 4.
8. For the *dimensions*, we opted for a distance of 70 cm between adjacent input nodes and 60 cm between adjacent layers. This allowed us to fit a network with 50 input nodes in the space available at the venue.
9. For setting up the network, we used a *rope with markers* placed every 70 cm so that two helpers were stretching the rope across the area and several helpers could place the ceramic tiles along the rope; once done, the rope was moved to the next layer, and this was repeated until all layers where set up. This way, all tiles could be placed within an hour of teamwork. A megaphone proved to be useful for coordination.

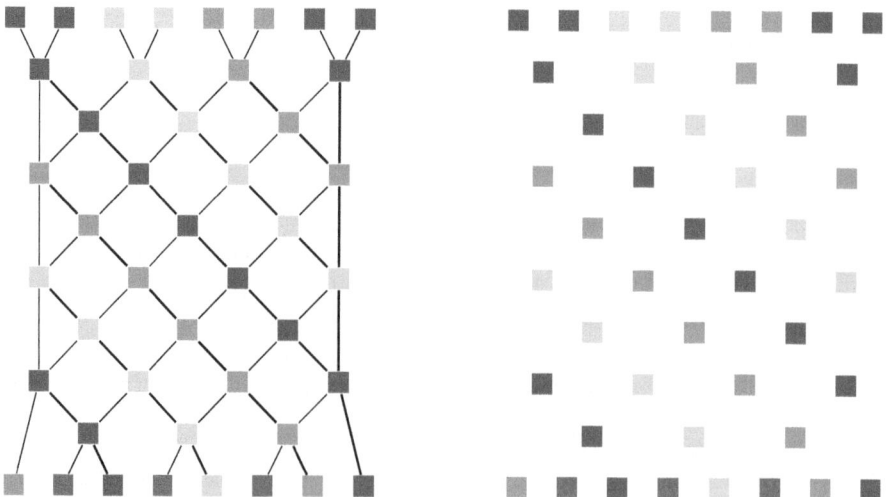

Fig. 4. Left: Odd-even transposition network with 8 input nodes and color coding; input nodes are on the top, output nodes on the bottom, and edges are directed top-down. Right: the same network drawn without edges.

We had the tiles delivered the day before the event. On the day of the event, a small team started tracing out the network outline and preparing the measuring rope. When the students arrived, we started to place the tiles. The entire setup was accomplished within about two hours. Then, we gave some instructions on the sorting process and did some demonstrations on a small part of the network with edges indicated by chalk. We already had some instruction sessions with the students at their schools a few days prior. Then, we proceeded to the actual sorting with all 50 numbers.

The first two runs produced some mistakes; the third run went smoothly and reasonably fast. We repeated once more so that everybody could participate as a player at least once. On the reverse side of each sorting card, we printed a letter so that after the sorting was completed, we issued a command to all players to reveal the reverse side of their cards, which displayed a message (see Fig. 3).

Although this activity retains the learning objectives of the original HSN activity, the increased scale of the network provides an excellent basis for discussing the topic of algorithm efficiency with the students. In addition to the educational purpose, a giant HSN activity opens an excellent channel for communicating the topic of algorithms to the general public.

For documentation, we had a photographer and a videographer team present, who captured the setting up of the network and the sorting activity. A video documentation is available at YouTube[2] and Zenodo[3]. We consider the event successful based on the positive feedback from the participating students and their teachers; for any similar events in the future, we recommend a more systematic gathering of feedback with questionnaires. Since the event was featured in several Austrian newspapers, it helped to raise public awareness on the importance of the subject.

4 Parallel HSNs

After organizing several events with participants from different age groups, using a 6-input HSN, we observed that a race character can improve the overall experience and engagement. The participants are divided into groups of six, then the time is recorded for each group completing the sorting, the fastest group wins.

We realized that performing the race in parallel with two sorting networks is more fun so that time-tracking is not necessary. This way, one can have a tournament where pairs of groups compete with each other, while retaining the learning objectives of the single-network HSN activity.

To this aim, we replaced our improvised sorting network drawn with colored tape on a tarp with two identical networks professionally printed. We found a company specialized in printing large advertisement banners. Banners of width 310cm and arbitrary length can be ordered.

We designed several variants of sorting networks (with minimal size and depth) as shown in Fig. 5, with 6–8 input nodes. We settled on the one with 6

---

[2] https://youtu.be/9KGRmKPdeoU.
[3] https://doi.org/10.5281/zenodo.13283701.

input nodes. Minor and major edges are distinguished by color and by the width of the line; it is easy to memorize "smaller number—thin line, larger number—thick line."

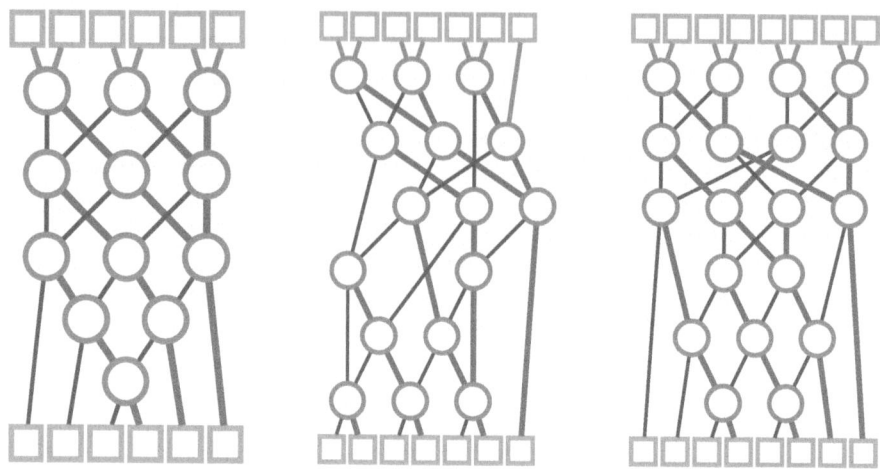

Fig. 5. Templates for HSNs with 6, 7 and 8 input nodes.

We note that the logical structure of a sorting network does not determine the way it is drawn. In fact, it is a nontrivial task to find a drawing that, for instance, minimizes the number of crossings or the distance between adjacent nodes. For a drawing of a sorting network one would require that all input nodes, output nodes, and comparator nodes of the same level, are placed along a line, respectively. But only the output nodes are required to appear in sorted progression, for the other nodes, any permutation is admissible, hence the total number of possibilities is exponential.

We organized HSN races with the two parallel networks on several occasions. Recently, the parallel networks were part of an activity at the *FPT Fest*[4] in Bergen, Norway. Since most of the participants were researchers in parameterized complexity [5], we printed parameterized complexity classes

$$\genfrac{}{}{0pt}{}{\text{polynomial}}{\text{time}} \subseteq \genfrac{}{}{0pt}{}{\text{polynomial kernel}}{\text{\& decidable}} \subseteq \text{FPT} \subseteq W[1] \subseteq W[P] \subseteq XP,$$

linearly ordered by set inclusion, on the sorting cards. This is just one example on how an HSN activity can be adjusted to a special group of participants (Fig. 6).

---

[4] FPT Fest 2023 in the honour of Mike Fellows, June 12–16, 2023, https://www.uib.no/en/rg/algo/160260/fpt-fest-2023-honour-mike-fellows.

Fig. 6. Mike Fellows conducts a sorting race with two parallel sorting networks at the FPT Fest in Bergen, 2023.

5 Future Work

As discussed above, we chose the simple odd-even transposition sorting network for the giant HSN because of its grid-like layout, so an edgeless drawing was still working. This choice is somewhat unsatisfactory from the computer science perspective, given the existence of significantly more efficient sorting networks of smaller depth and size (see Sect. 2). Therefore, we propose an approach that allows us to use more efficient sorting networks for a large HSN based on an edgeless drawing for future HSN events.

The idea is to place the comparator nodes on a grid and annotate them with coordinates (ℓ, i) where ℓ is the level and i the index within the level, and with the coordinates of the successor nodes via the minor and the major edge, (ℓ', i') and (ℓ'', i''), respectively, possibly drawn as follows:

$$\boxed{\begin{array}{c} \ell, i \\ \ell', i' < \ell'', i'' \end{array}}$$

Output nodes are referred to by their index. Figure 7 shows an example of a sorting network with 8 input nodes with the smallest possible depth of 6. Already here, the reduction in depth by 2 compared to the odd-even transposition sorting network, as displayed in Fig. 4, will compensate for the more difficult task for the

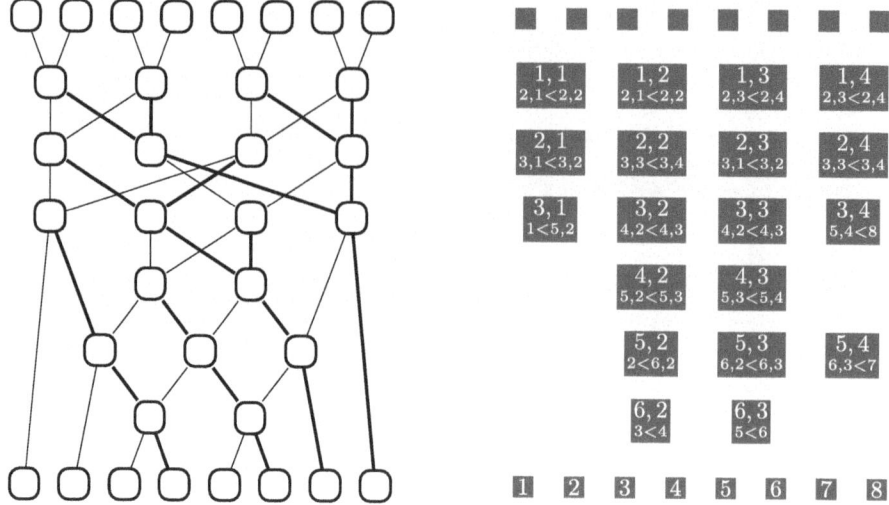

Fig. 7. Left: an optimal sorting network for 8 input nodes. Right: an edgeless drawing of the network using annotations.

players to find and proceed to the next node; for a larger number of inputs, this reduction is even more pronounced.

Indeed, it might be exciting to have in parallel two large sorting networks with, say, 50 input nodes each, one based on the odd-even transposition sorting and one based on Batcher's bionic sorting, and to conduct a race. This could provide a strong educational effect on the value of designing efficient algorithms.

For the drawing of a given sorting network that works well as an HSN, it is essential to keep the edges short, so that during the sorting, the players do not need to cover long distances when moving from one node to the next. As mentioned above, finding such a drawing is an interesting optimization problem on its own, and could be solved together with the participating students at workshops to be held before the actual sorting event.

6 Conclusion

The extensions to the Human Sorting Network (HSN) activity presented in this paper demonstrate the versatility and scalability of this educational tool. The large-scale 50-input HSN successfully engaged high school students in a collaborative, full-day event that brought algorithmic concepts to life on a grand scale. This implementation showcased the potential for adapting sorting networks to create impactful public events that promote computer science awareness.

The parallel HSN setup, on the other hand, introduces a competitive element that enhances engagement and allows for tournament-style activities. This variant has proven effective across various age groups and can be easily customized for specific audiences, as demonstrated by its use with parameterized complexity classes at a research conference. Both extensions offer valuable lessons for educators and researchers looking to expand the reach and impact of CS Unplugged activities. These innovations in HSN activities offer several benefits:

1. They provide hands-on experience with algorithmic concepts for a wide range of participants, from children to adults.
2. They foster teamwork and collaborative problem-solving.
3. They can be adapted to different contexts and knowledge levels.
4. They create opportunities for public outreach and raising awareness about computer science concepts.

By sharing our experiences and methodologies, we hope to inspire further innovations in hands-on computer science education and contribute to the ongoing effort to make algorithmic thinking accessible and engaging for diverse audiences.

Future work could explore more efficient network layouts for large-scale HSNs and investigate the educational outcomes of these extended activities compared to the traditional HSN setup.

Acknowledgments. The core team for the large HSN project comprised Stefan Szeider (idea, scientific direction), Agata Ciabattoni (scientific advice), and Mihaela Rozman (coordination). The core team is grateful to Hermann Morgenbesser (Future Learning, PH Wien, and the International School Klosterneuburg), Denise Hackner and Bernhard Klimbacher (Karl-Popper School Vienna), as well as Philipp Prinzinger (EduLab of TU Wien) for their help and acknowledge the funding by the Vienna Business Agency and the Austrian Ministry for Education (BMVIT).

References

1. Ajtai, M., Komlós, J., Szemerédi, E.: An $O(n \log n)$ sorting network. In: Johnson, D.S., et al. (eds.) Proceedings of the 15th Annual ACM Symposium on Theory of Computing, 25-27 April, 1983, Boston, Massachusetts, USA, pp. 1–9. ACM (1983). https://doi.org/10.1145/800061.808726
2. Batcher, K.E.: Sorting networks and their applications. In: American Federation of Information Processing Societies: AFIPS Conference Proceedings: 1968 Spring Joint Computer Conference, Atlantic City, NJ, USA, 30 April - 2 May 1968. AFIPS Conference Proceedings, vol. 32, pp. 307–314. Thomson Book Company, Washington D.C. (1968). https://doi.org/10.1145/1468075.1468121, https://doi.org/10.1145/1468075.1468121
3. Bell, T., Rosamond, F., Casey, N.: Computer science unplugged and related projects in math and computer science popularization. In: Bodlaender, H.L., Downey, R., Fomin, F.V., Marx, D. (eds.) The Multivariate Algorithmic Revolution and Beyond. LNCS, vol. 7370, pp. 398–456. Springer, Heidelberg (2012). https://doi.org/10.1007/978-3-642-30891-8_18
4. Bell, T., Witten, I.H., Fellows, M.: Computer Science Unplugged: Off-line activities and games for all ages (original book) (1999). http://csunplugged.org
5. Downey, R.G., Fellows, M.R.: Fundamentals of parameterized complexity. Texts in Computer Science. Springer (2013)

6. Karpinski, M.: Encoding cardinality constraints using standard encoding of generalized selection networks preserves arc-consistency. Theoretical Comput. Sci. **707**, 77–81 (2018). https://doi.org/10.1016/J.TCS.2017.09.036
7. Knuth, D.E.: The Art of Computer Programming, volume III: Sorting and Searching. Addison-Wesley (1973)
8. Parberry, I.: Parallel complexity theory. Research notes in theoretical computer science, Pitman (1987)
9. Paterson, M.: Improved sorting networks with $o(\log n)$ depth. Algorithmica **5**(1), 65–92 (1990). https://doi.org/10.1007/BF01840378
10. Seiferas, J.I.: Sorting networks of logarithmic depth, further simplified. Algorithmica **53**(3), 374–384 (2009). https://doi.org/10.1007/S00453-007-9025-6

Open Access This chapter is licensed under the terms of the Creative Commons Attribution 4.0 International License (http://creativecommons.org/licenses/by/4.0/), which permits use, sharing, adaptation, distribution and reproduction in any medium or format, as long as you give appropriate credit to the original author(s) and the source, provide a link to the Creative Commons license and indicate if changes were made.

The images or other third party material in this chapter are included in the chapter's Creative Commons license, unless indicated otherwise in a credit line to the material. If material is not included in the chapter's Creative Commons license and your intended use is not permitted by statutory regulation or exceeds the permitted use, you will need to obtain permission directly from the copyright holder.

Distance Teaching of Mathematical and Computer Disciplines During the War in Ukraine

Galina Bulanchuk[1], Oleh Bulanchuk[2](✉), Olena Piatykop[1], and Valentyna Ilkevych[3]

[1] SHEI "Pryazovskyi State Technical University", Gogolya Street 29, Dnipro 49000, Ukraine

[2] Junior Academy of Sciences of Ukraine, Dehtiarivska Street 38-44, Kyiv 04119, Ukraine
ol_bulanch@ukr.net

[3] Separate Structural Unit «Mariupol Polytechnic Professional College of the SHEI "Pryazovskyi State Technical University"», Gogolya Street 29, Dnipro 49000, Ukraine

Abstract. The experience of distance learning during COVID-19 has become invaluable for continuing lifelong learning process under the war conditions in Ukraine. That experience was especially helpful for the survival of universities relocated from the occupied territories, where distance learning became the only possible format. The main problem for Ukrainian universities during the war is that students often cannot attend online meetings due to poor communication, blackouts, or air alarms. The task is to organize the educational process to ensure its effectiveness. Ideally, the teacher should create a learning environment accessible from anywhere in the world at any time convenient for the student.

This paper analyzes and systematizes the authors' experience in teaching mathematical and computer disciplines remotely during a full-scale invasion. We have identified four main criteria that a course should satisfy: completeness, self-sufficiency, finality, and relevance. The teacher must create high-quality course content that is accessible for students' independent learning. This includes creating a syllabus, preparing and recording short video lectures, conducting consultations, organizing control activities (such as forming tests and assignments), and holding retrospectives where the author's solutions are reviewed and analyzed, especially for computer disciplines and programming.

The authors believe that this model of interaction between the teacher and students during the war is key, as it allows students, in addition to communicating directly with the teacher in class, to have well-prepared and structured material for independent work and quick feedback from the teacher through messengers.

Keywords: distance learning · wartime education · mathematics · computer science · online learning · educational technology

1 Introduction

Distance learning dates back to the late 1800s. This happened through correspondent training. The student could send his work to the teacher and receive the teacher's comments and textbooks by mail. The history of distance learning is described in [1–3]. Distance learning is primarily online learning with a teacher in synchronous mode via the Internet. According to the studies presented in [3], in 2012, 69% of major academic universities indicated that online learning was critical to their long-term strategy, and of the 20.6 million students enrolled in higher education, 6.7 million were enrolled in an online course. Trends and developments in the field of distance education research during 2009–2013 are described in [4].

During the COVID-19 pandemic, more than 1.7 billion students worldwide have stopped in-person learning due to the spread of the coronavirus disease. This has become a huge challenge for teachers, who urgently and must massively adapt all their classes to distance learning in order to maintain continuity of education with the same quality. The development of new strategies for organizing distance learning has begun. Paper [5] presents feedback from 20 students and teachers who participated in a survey during quarantine. They admit that in two months of practical work they learned more about distance education than in the last 10 years. It should be noted that the transition to online learning during COVID-19 caused some discomfort among students. The work [6] assessed students' attitudes towards the rapid transition to fully online learning. Most students responded that online learning was not the same as in-class learning (91.5%). Many students (75.6%) responded that they experienced some level of anxiety about the rapid transition to online. A follow-up survey after 3 weeks showed improved attitudes among students who said they were less worried about online learning.

Research has also been conducted on students' attitudes toward distance learning in mathematics [7]. Research showed that students had mixed feelings about their distance learning experiences. The most negative attitudes were towards the lack of interaction with teachers and colleagues and the need to spend long periods of time in front of a screen. In addition, 78.3% of participants did not see an advantage in choosing a distance learning mode if such a choice were possible.

In Ukraine, due to the spread of COVID19, a long-term quarantine was introduced in 2020, which led to the transition to distance learning in schools and universities. Over the next two years, the learning process took place in a mixed format. During all this time, teachers have gained significant experience in conducting online classes. This experience became invaluable for continuing the continuous learning process in war conditions when, on February 24, 2022, the Russian Federation launched a full-scale invasion of Ukrainian territory. This experience especially helped the survival of universities displaced from the occupied territories, for which distance learning became the only possible format in such conditions. Now, in addition to everything else, for students and teachers in Ukraine the situation is further complicated by frequent air raid alerts and blackouts. This experience is somewhat unique in the world. For example, in Israel, despite the war, children and students study full-time. The basis of security in Israel is the Civil Defense Law of 1951, which obliges all types of buildings to be equipped with shelters and bomb shelters. The experience of Ukrainian teachers can be useful both in case of war and in peacetime. Now it is just beginning to accumulate and generalize (see, for example, works [8,9]).

It should be noted that during the war, three categories of students were formed in Ukraine: those in unoccupied territories, those in occupied territories and those abroad. Each of these categories has its own problems and challenges. For relocated universities, the division into these categories also applies to teachers. At the same time, the task is to organize the educational process in such a way as to ensure its effectiveness for all categories of students. Of course, a key role in such conditions will be played not only by the qualifications of the teacher, but also by his knowledge of information educational technologies. Ideally, one should strive to create a learning environment that will be accessible from anywhere in the world at any time convenient for the student.

This paper analyzes and systematizes the authors' experience in teaching mathematical and computer disciplines remotely during the period of a full-scale invasion.

2 Distance Learning Methodology

The main problem of Ukrainian universities during the war is that, objectively, students cannot always attend online meetings. For students who are in Ukraine, this is due to air raid alerts and blackouts. For students in the occupied territories - mainly due to poor communications. This problem also affects students who are abroad, since most of them attend integration courses or have to work. Therefore, the teacher's main attention should be focused on preparing the material and tasks in such a way that it is available to students when they have all the conditions for this (light, Internet and no air raid). In this case, an important factor is also the brevity of the prepared material. Of course, communication with a teacher does not lose its value, but in such conditions, in addition to online classes, it occurs individually or in groups through different messengers: Viber, Telegram, WhatsApp, etc. Provided that high-quality material is prepared for students, the need for such individual consultations via the messenger will be minimized and will not be burdensome for the teacher.

The research findings are derived from an analysis of distance learning across nine groups, totaling 113 students, as well as from the authors' own online courses and the completion of approximately 30 online courses on platforms such as edX, Coursera, Prometheus, DataCamp, Stepic, ulearn.me, and courses provided by SoftServe, Ciklum, GoIt, and QATestLab. Through an evaluation of student performance, survey responses, and personal experience with various online courses, we have identified four primary criteria that a well-designed course should meet to ensure effectiveness. These criteria are not only applicable to mathematical or computer science disciplines but are also relevant to other fields of study and in-person learning environments. Our analysis indicates that the absence of any of these criteria can significantly diminish the effectiveness of the training. The key criteria for a quality course are as follows:

- Completeness. The teacher should determine whether all theoretical information is covered by tests or assignments. Theoretical information that is not covered by tests or assignments should not be provided, as it overloads cognitive perception and discourages the desire to learn. In this case, there is a certain analogy with software testing, where there is a requirement for automated tests to cover the program's code. Some online courses suffer from inadequately linking theoretical information with practical tasks and tests. In some instances, the theoretical content does not facilitate the resolution of practical problems, leading to a disconnect between theory and practice. There are even cases where the theoretical material provided misguides learners. Another issue arises when the theoretical content is insufficient or presented in a manner that hinders the generalization and adaptation of the knowledge to solve modified problems.
- Self-sufficiency (closure). One should ask themselves whether the provided knowledge is sufficient to complete the assigned tasks. The necessity to use other sources significantly increases the learning time and can lead to frustration due to the large volume of information that needs to be processed. It may be unclear to the student what exactly is relevant for mastering the course. It should be noted that during wartime, resources such as the time of both the instructor and the student should be carefully considered. This is especially true considering the mental and psychological state during air alarms or blackouts. Real time available for completing tasks becomes less. Additionally, students and instructors can become frustrated when they feel this time is being wasted. This is particularly relevant given that many students are forced to work to support themselves or their families. As a result, students may not have enough time to complete their tasks within the allocated time. Therefore, providing additional references should be avoided as it can be detrimental and distracting. As an exception, references can be provided, but then it is necessary to specify what to pay attention to and also prepare a test on these questions.
- Finality. It is desirable that courses, where appropriate, conclude with a creative task or project that can be added to one's portfolio. This will help in finding future employment. In this case, requirements for the project should be formulated. The degree of detail can vary, but the requirements should not be overly strict. The project should be designed so that students utilize most of the knowledge and skills covered in the course. External references are useful here, and it is additionally desirable to specify what exactly can be used. This is necessary so that there are prompts during the project work using materials and ideas. Recommendations for the internal implementation of the project should only be advisory and not mandatory. An important factor in improving the implementation of a project is the continuous monitoring of intermediate results and facilitating the exchange of experiences among

students. In the context of computer science, employing Scrum methodology, with its structured meetings such as planning sessions and retrospectives, proves to be highly effective.
- Relevance. The information provided in the course should correspond to the current state of the discipline being studied (especially relevant to computer disciplines that utilize frameworks and libraries that are constantly updated).

Among the materials we filled our courses with, we used the following types:

- Video story (lecture). It's not just a recorded Zoom video conference. It is advisable to make separate short (up to 15 min) videos on questions that will then be on the exam. For example, we published videos on higher mathematics on the YouTube channel and provided links in the materials for a separate lecture, which we posted in Google Classroom. There were usually several such videos for the lecture (3–4 videos). The student can look at exactly those questions that he does not understand. When these videos are first filmed, it takes a long time (sometimes several takes). However, this material can be used in subsequent years for another group for those disciplines where fundamental changes in content do not occur (for example, mathematics).
- Video story with test inserts.
- A video story asking you to pause the lecture and do the formulated task yourself, and then watch how the author does it.
- Single-choice tests.
- Multiple choice tests.
- Checked tasks (tasks) (the contents of the file with errors are displayed).
- Task(s) that are checked automatically and hints (points are deducted when using such hints).
- Problem(s) checked by teachers or teaching assistants.
- Project checked by the teacher.
- Project not reviewed by instructor.
- A project tested by several students in this course.

As an example in Fig. 1 shows the material for the lecture "Primary. Indefinite integral." [10] It includes a video explanation (up to 5 min) of individual questions (initial, tabular integrals and inclusion under differential) on the YouTube channel and lecture notes in pdf format.

Effective and efficient learning significantly benefits from the use of lecture notes. In our course on higher mathematics, we utilized the Zoom platform [11], a Galaxy Tab S6 Light tablet with an S Pen for handwritten mathematical formulas, and the Microsoft OneNote program [12] for cloud storage of notes. The advantage of using OneNote lies in its ability to systematically organize all lectures and practical exercises, enabling the provision of a comprehensive link to students for exam preparation. Furthermore, lecture notes can be distributed in PDF format after each session. Practical exercises can be conducted synchronously within this platform, and even the free version offers a multitude of features. The Fig. 2 illustrates the course structure and provides an example of

Fig. 1. Fragment of video materials for the lecture.

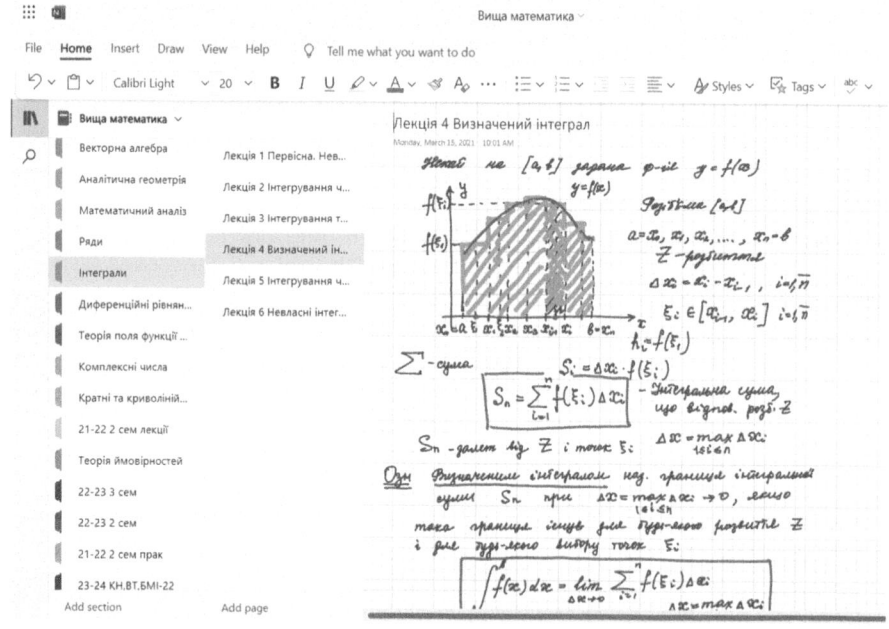

Fig. 2. Structure of the Higher Mathematics course in OneNote.

a lecture written on a tablet. The instructor can access this environment simultaneously from multiple devices, such as a computer and a tablet, if necessary. Additionally, online boards such as CleverMaths [13], Miro [14], and Jamboard [15] can be utilized to further facilitate the learning process.

In our opinion, tests play a crucial role in distance learning not only for verification but also for knowledge formation. To effectively form knowledge, we utilized tests with multiple correct answers, typically offering 6–7 options, and allowed students to take these tests multiple times (more than five attempts). These tests were structured to cover all theoretical material, essentially serving as a guide for studying the theory and were taken in a specific sequence rather than randomly. Analysis of students' material assimilation showed that the quality of knowledge in the control group was 23% higher compared to traditional study methods. Tests can be classified according to different criteria, with two main characteristics being the number of attempts (repeatability) and the test execution time. Let's highlight the main types of tests. By test repeatability:

– once (usually finals and during lecture);
– two times (current in the lesson);
– training (10 or more attempts).

Time to complete tests:

– unlimited time;
– limited time for the entire test;
– limited time per question.

From the perspective of quality control of material comprehension, the best approach is timed tests for each question. In conditions of internet or electricity outages, it is advisable to organize them so that each answer is submitted separately. This way, in case of an outage, no more than one answer would be lost, and testing can be resumed with the previously submitted answers taken into account. Analysis of online courses has shown that in most cases, the following testing methods are used:

– At the beginning (or end) of a live lecture, a short quiz with a single answer option is conducted (for example, in the Kahoot system [16]).
– Testing embedded within recorded lectures.
– Training tests (up to 10 attempts) with multiple-choice questions, with a time limit for the entire test (the test with the best result is counted).
– The final test is taken once and has a time limit.

In our opinion, training tests and the final test at the end of each section or the entire course yield the greatest effect. In this case, while preparing for the final test, students have the opportunity to review all training tests once more. Clearly, such repetition allows for the formation of more stable neural connections.

Homework can be classified as follows:

– Problems that are checked automatically; the number of attempts is unlimited by the time of completion.
– Tasks checked by the teacher, when by the time of submission the student can submit the work for checking more than once and receive several reviews.
– In the case of programming, tasks are provided in the form of an archive containing code samples and automated tests. Task conditions are formulated through requirements in a separate file. It is also convenient to use Github Classroom for this purpose.

The most effective in terms of student time and learning organization is verification by the teacher. This approach instills several useful habits in students:

- tasks are completed on the day they are assigned (to receive feedback quickly and have time to get the next one; typically, up to three feedbacks can be received);
- it counteracts procrastination (the first version of the answer does not have to be perfect but must be quick);
- it familiarizes students with the Scrum methodology (where the result is achieved in several iterations).

However, this approach increases the teacher's workload and requires more precise formalization of tasks (where not only the final result is evaluated, but also the intermediate steps).

If there is a project, it is typical for the teacher to check the project with the provision of comments or a retrospective (in class, with the consent of the student, his project is analyzed).

Consultations are organized as follows: a chat is created where students can exchange information and receive advice from the teacher. In the chat it is forbidden to give direct answers, only advice and tips.

Where lectures and practical classes are taught by one teacher, combined lecture-practical classes should be done, which are easier for students to learn. Long lectures cause a decrease in students' attention and motivation, because it is difficult to concentrate on a significant amount of information for a long time.

Another essential aspect of a high-quality online course is the ability to quickly review material. In this regard, flashcards are particularly effective, as they enable swift and efficient revision of previously covered content. In this case, memory cards are very effective, which allow you to quickly view and update the passed material in memory.

It is interesting to note that the principles of building quality online courses align well with John Biggs' educational theory of Constructive Alignment [17], which aims to enhance the effectiveness of teaching and learning in higher education. Constructive Alignment emphasizes active engagement, where students are active participants in their learning journey, engaging with and reflecting on what they are learning. It also highlights the importance of clear objectives, ensuring that learning outcomes are explicitly stated and that teaching activities and assessments are designed to directly support these outcomes.

In designing such courses, the concept of Backward Design, as articulated by Wiggins and McTighe [18], should be employed. This approach begins with the end goals in mind, involving the identification of desired learning outcomes first, followed by the planning of instructional methods and assessments to achieve those outcomes. By focusing on the desired results and working backward, educators can create a more coherent and effective learning experience that aligns with both Constructive Alignment and the principles of Backward Design.

3 Conclusions

Online learning provides opportunities to optimize the educational process, thereby increasing its effectiveness. Learning can be deemed effective if knowledge is well-mastered and skills are developed with minimal time investment from the student, although this demands greater qualifications and effort from the teacher. This is clearly a challenging task. The teacher's role extends beyond the art of teaching to include developing course content, creating programs, preparing and recording video lectures, conducting consultations, and organizing assessments (including tests, assignments, and retrospectives).

Experience has shown that the most effective forms of online learning during wartime include short video lectures, quizzes during lectures, video lectures with pauses for self-completion tasks, timed training tests covering the entire material (primarily multiple-choice questions), assignments reviewed multiple times by the instructor with critical feedback within a limited timeframe, automated error-checking, a final test derived from training tests, and a summative assessment involving a final project, which can be added to a portfolio.

The authors believe that this model of teacher-student interaction during wartime is crucial. It allows students to have well-prepared and structured material for independent work, alongside the benefit of direct communication with the teacher during class and quick feedback via messaging platforms.

References

1. Banas, E.J., Emory, W.F.: History and issues of distance learning. Public Adm. Q. **22**(3), 365–383 (1998)
2. Harting, K., Erthal, M.: History of distance learning. Inf. Technol. Learn. Perform. J. **23**(1), 35–44 (2005)
3. Kentnor, H.: Distance education and the evolution of online learning in the united states. Curr. Teach. Dialogue. Volume 17, **1**(2), 21–34 (2015)
4. Bozkurt, A., et al.: Trends in distance education research: a content analysis of journals 2009–2013. Int. Rev. Res. Open Distrib. Learn. **16**(1), 330–363 (2015). https://doi.org/10.19173/irrodl.v16i1.1953
5. Dietrich, N., Kentheswaran, K., et al. : Attempts, Successes, and Failures of Distance Learning in the Time of COVID-19. J. Chem. Educ., American Chem. Society, Division of Chemical Education. **97**(9), 2448–2457 (2020)
6. Unger, S., Meiran, W.: Student attitudes towards online education during the COVID-19 viral outbreak of 2020: distance learning in a time of social distance. Int. J. Technol. Educ. Sci. (IJTES). **4**(4), 256–266 (2020)
7. Almarashdi, H., Jarrah, A.: Mathematics distance learning amid the COVID-19 pandemic in the UAE: high school students' perspectives. Int. J. Learn. Teach. Educ. Res. **20**(1), 292–307 (2021)
8. Ivanyuk, I.V.: Features of distance learning during the war in 2023. In: Digital tools for the development of information literacy and critical thinking of students: a collection of webinar materials as part of the Fifteenth International Exhibition "Innovation in Modern Education" on October 26, 2023 (Kyiv, October 26, 2023). ICO NAPN of Ukraine, Kiev, Ukraine, pp. 20–24 (2023). https://lib.iitta.gov.ua/737644/
9. Khaniukov, O., Smolianova, O., Shchukina, O.: Distance learning during the war in Ukraine: experience of internal medicine department (organisation and challenges). Art Med. **3**. 134–138. (2022). https://doi.org/10.21802/artm.2022.3.23.134
10. Higher mathematics Homepage. https://youtube.com/@high_mathematics?si=TJlKxS444eYOphiA

11. One platform to connect | Zoom Homepage. https://zoom.us/. Accessed 23 Jun 2024
12. Onenote Homepage. https://www.onenote.com/. Accessed 23 Jun 2024
13. Download CleverMaths by Clevertouch Homepage. https://clevermaths.software.informer.com/download/. Accessed 23 Jun 2024
14. Miro | The Visual Workspace for Innovation Homepage. https://miro.com/. Accessed 23 Jun 2024
15. Google Jamboard: Whiteboard App for Business | Google Workspace Homepage. https://workspace.google.com/products/jamboard/. Accessed 23 Jun 2024
16. Kahoot! | Learning games | Make learning awesome! Homepage. https://kahoot.com/. Accessed 23 Jun 2024
17. Biggs, J.: Constructive alignment in university teaching. HERDSA Rev. High. Educ. **1**, 5–22 (2014)
18. Wiggins, G., McTighe, J.: Understanding by Design. ASCD (2005)

Author Index

B
Bell, Tim 130
Bilbao, Javier 168
Bulanchuk, Galina 205
Bulanchuk, Oleh 205

D
Dagienė, Valentina 3, 168
Desikachari, Krishnamachari 66

F
Fellows, Michael R. 15

G
Gülbahar, Yasemin 168

H
Hickman, Henry 130
Hromkovič, Juraj 136

I
Ilkevych, Valentyna 205

J
Jacob, Riko 39
Jain, Manish 162

K
Kaarto, Heidi 168
Kilpi, Janica 168

L
Laakso, Mikko-Jussi 168
Lacher, Regula 136
Landman, Martina 50
Laplante, Sophie 77
Lehner, Lukas 50

Löffler, Maarten 104
Lütscher, Pascal 93

M
Müller, Matthias 93

N
Nitschke, Łukasz 185

P
Parviainen, Marika 168
Pears, Arnold 168
Perekietka, Paweł 185
Perez, Loris 77
Piatykop, Olena 205
Pluhár, Zsuzsa 168

R
Ramanujam, R. 117
Ren, Michał 185
Rosamond, Frances A. 15

S
Shah, Vipul 117
Silvestri, Francesco 39
Sysło, Maciej M. 147
Szeider, Stefan 194

T
Thakkar, Jay 162
Thamburaj, Robinson 66
Thomas, Gnanaraj 66
Tissot, Sylvie 77

V
Vettier, Lou 77

W
Weißing, Benjamin 93

© The Editor(s) (if applicable) and The Author(s), under exclusive license
to Springer Nature Switzerland AG 2025
H. Fernau et al. (Eds.): CMSC 2024, LNCS 15229, p. 215, 2025.
https://doi.org/10.1007/978-3-031-73257-7

SPRINGER NATURE

GPSR Compliance

The European Union's (EU) General Product Safety Regulation (GPSR) is a set of rules that requires consumer products to be safe and our obligations to ensure this.

If you have any concerns about our products, you can contact us on ProductSafety@springernature.com

In case Publisher is established outside the EU, the EU authorized representative is:

Springer Nature Customer Service Center GmbH
Europaplatz 3
69115 Heidelberg, Germany

The manufacturer's authorised representative in the EU is Springer Nature Customer Service Centre GmbH, Europaplatz 3, 69115 Heidelberg, Germany. If you have any concerns regarding our products, please contact ProductSafety@springernature.com

Printed and bound by CPI Group (UK) Ltd, Croydon, CR0 4YY

26/03/2026

02078962-0003